Process Safety 2024

Held at the 2024 AIChE Spring Meeting and 20th Global Congress on Process Safety

New Orleans, Louisiana, USA
24-28 March 2024

ISBN: 979-8-3313-1719-5

Printed from e-media with permission by:

Curran Associates, Inc.
57 Morehouse Lane
Red Hook, NY 12571

Some format issues inherent in the e-media version may also appear in this print version.

AIChE
120 Wall Street, FL 23
New York, NY 10005-4020

Phone: (800) 242-4363
Fax: (203) 775-5177

www.aiche.org

Additional copies of this publication are available from:

Curran Associates, Inc.
57 Morehouse Lane
Red Hook, NY 12571 USA
Phone: 845-758-0400
Fax: 845-758-2633
Email: curran@proceedings.com
Web: www.proceedings.com

.TABLE OF CONTENTS

LOOKING TOWARDS THE FUTURE: PROCESS SAFETY AND SUSTAINABILITY

92c Wells Safety Critical Equipment Integrated Risk and Barrier Management Tool

Natalia Bellesi, Vishesh Kapur Sr.

One of the key challenges in Process Safety Management is the assurance of Barrier effectiveness in the field. In particular for Wells activities where multiple contractors are involved, efficient coordination of all interfaces and activity stages is critical in order to validate Safety Critical equipment (SCE) and barriers are well maintained and effectively operating. Taking into account this goal we designed a new tool for handling Safety Critical Equipment in Shales Wells Completion area. This tool is being used in Shell Argentina and Canada Shales Assets.

While shales drilling activities utilize same contractors and packaged equipment, Completion work is done by several different contractors and equipment. Furthermore, one contractor can bring different equipment for each Completions activity/service. This introduces complexity in handling Safety Critical Equipment Preventive Maintenance certification because each piece of equipment needs to be certified prior to completions start up. In order to provide this certification, contractors used to send emails with documents that assure all required maintenance was done but this adhoc method of email approval was not available in a centralized location to verify and visualize SCE. This resulted in difficulty to audit the process and manage risk adequately.

In order to develop this tool, first the Safety Critical Equipment list and associated tasks were reviewed based on bowtie scenarios, identifying key primary and secondary hardware barriers. Once the list was validated, a system was designed to manage the interphase between contractors loading Safety Critical Equipment information and field supervisor validating the information and authorizing the startup of activities. At this point, one of the risks we needed to consider was cybersecurity one in order to allow access to third parties to a Shell system without generating a potential IT risk. Then a visual platform was developed to create an easily accessible dashboard where all required stakeholders could check online the status of Safety Critical Equipment.

The implementation of this new system allows us to simplify the way Safety Critical Equipment PM is certified, providing visibility of SCE assurance, and consolidating a user friendly and auditable platform to follow up on barrier health. Also, the improved visualization increases the understanding of SCE importance as critical barriers to reduce the probability of an incident and gain accountability of field roles in identifying their responsibilities for keeping these barriers effective in the operation. Furthermore, this tool has the capability of assuring the personnel operating the equipment have required skillset/training. This will help manage the personnel safety risk.

The tool can be accessed via various devices such as Ipad, Tablet, mobile phone, etc. This provides flexibility for the hardware being used to open the application. Furthermore, this application has offline capabilities in which documents can be uploaded without network connectivity.

In conclusion, this Fit for Purpose tool optimized the Ways of Working by streamlining the assurance and approval process thereby strengthening this Process Safety culture in the Wells and contractor organization which will drive/ achieve sustainable business outcomes.

92d Integrated PHA Approach for an Auditable Barrier Management Regime

Khama Matiti

The Oil and Gas Industry is constantly evolving and increasing in complexity. Operators are seeking new strategies to increase the auditability and connectivity between disciplines and their studies. Barrier management is a developing methodology that is utilised, to systematically establish and maintain barriers so that the risk faced at any given time can be handled. This is achieved by preventing undesirable incidents from occurring or by limiting the consequences should such accidents occur. Barrier management includes the processes, systems, solutions, measures and tools which must be in place to ensure the necessary risk reduction through its implementation. Barrier management establishes an easily communicated risk picture.

Companies are having to look at streamlining methodologies to ensure effective and efficient barrier management techniques.

A holistic barrier management involves the a), the identification of hazards and their barriers, the determination of barrier performance standards, and the monitoring of barrier performance during operations and their continued applicability in an ever changing operational environment. There are various techniques and tools available to safety professionals when considering overall barrier management. However, despite some efforts in the past, there is still no standard methodology on how to practically combine these complex standalone techniques into one easily understood combined barrier model. Techniques such as HAZOP, LOPA, RBI, Bowtie, Alarm Rationalisation, Prestart-up safety audits, and human factors studies are traditionally used to address concerns due to design issues, mechanical integrity issues, process and operations issues and external events. However, these studies are performed in isolation and are rarely combined into one distilled easily understood risk picture. This results in differing competing conclusions from the various studies and a complex risk picture which is not easy for frontlines operations staff to understand and implement.

To address the potential issues of performing these studies in isolation and to ensure the safe operation of hazardous facilities, a formal, systematic and integrated system will be presented. As a sample case, an integrated HAZOP, LOPA approach was developed for a large existing onshore oil and gas facility to address process risks. This integrated approach is presented as an example to illustrate how complex studies such as HAZOP and LOPA were distilled into a pictorial and easy to use bowtie for operations. Bowtie diagrams are often used to communicate a holistic picture of barrier management efforts. They take into account all the various studies undertaken by the differing disciplines. The diagrams give a clear illustration of the relationships between the hazard, the causes of its realisation and its potential consequences. It includes easily understood pictorial descriptions of the control and recovery measures along with barrier degradation issues such as the effect that ageing has on a facility barrier management system.

Additionally, the paper will discuss examples techniques, which are being utilised or being considered for the integration of the degradation of barriers which are not deeply considered in HAZOP and LOPA, such as mechanical integrity, external event and design defects. When combined with the process risks discussed in HAZOP and LOPA a holistic unified risk picture can be presented e.g. the impact of omission of critical inspection and maintenance activities, human error during critical tasks operational activities, failure in ignition control, ineffective emergency response systems. The paper will therefore discuss the impact of deficient assurance activities, design, human factors and the overall process management system.

The paper will demonstrate how an integrated barrier management approach is currently being applied at an existing facility by focusing on integrating HAZOP, LOPA and Bowtie Studies. Potential future improvements to current practice will also be addressed. The paper will ensure that study findings, from various critical studies covering all potential failure mechanisms are clearly communicated via the pictorial bowtie, and provide a clear risk picture to practitioners, operations and management. This methodology results in a highly efficient and auditable workflow.

SPRING24
+20TH GCPS
A Joint AIChE and CCPS Meeting

Conducting PHA's in Companies with a Young PSM Culture

Guy Brouillard
Rio Tinto
Tour Deloitte,
1190 Avenue des Canadiens-de-Montréal,
Montréal, QC
H3C 1B3
Guy.brouillard@riotinto.com

Prepared for Presentation at
American Institute of Chemical Engineers
2024 Spring Meeting and 20th Global Congress on Process Safety
New Orleans, LA
March 24-28, 2024

Conducting PHA's in Companies with a Young PSM Culture

Guy Brouillard
Rio Tinto
Tour Deloitte,
1190 Avenue des Canadiens-de-Montréal,
Montréal, QC
H3C 1B3
guy.brouillard@riotinto.com

Keywords: Process Hazard Analysis, Culture, Training, Process Safety Information

Abstract

Process Hazard Analysis is a critical step in controlling high hazard processes. It is time consuming and requires significant resources. The review itself is normally only the tip of the iceberg.

To maximize the benefits from the safety review, it is essential that it is prepared in advance, and that the critical participants come ready for the review, and that they be available for the length of the review.

There are critical steps in the preparation, execution, and post review that mature PSM culture companies have experience with. If these are neglected, the quality of the review may be impacted.

This paper proposes tried and true steps that help maximize the return on investment in Process Hazard Analysis.

Process Hazard Analysis is a continually evolving domain. We are not conducting reviews in 2024 as we were in 1984. The following pages propose steps that will help improve the quality of the PHA.

1 Development of Hazard reviews

It is widely believed that the HAZOP (Hazard and Operability study) methodology was developed by the Imperial Chemical Industries (ICI) in the 1960's.

Since the development of the HAZOP, there have been many different methods designed to address process needs. Below is a short list of well-known methods in application today:

- HAZOP
- HAZID
- What-If
- Swift – Structured What-If
- What-If with Checklist
- FMECA (Failure Mode and Effect with Consequence Analysis
- Bow-Tie
- Fault Tree, Consequence Tree
- PFFM (Process Flow Failure Modes)
- QRA (Quantitative Risk Analysis)
- Etc

We will not go into the pros and cons of each method, but rather the common steps of planning, executing, and following up on the reviews themselves.

2 Planning

For best results, it is important to plan the review, and not conduct it add-hoc.

More and more companies have started using a PHA Charters to help them prepare for the review. It is important to have this charter approved by the person responsible for the resources that will participate in the review. Since this is a multidisciplinary endeavor, the approval often goes to the site senior leader. Some sites might do this differently. For projects, the approval is normally done by the project manager or their delegate.

2.1 Review Charter

2.1.1 Context of the review

Since there exists many different types of review, it is important to set the context.

Is this?

- A project PHA, what stage of the project is the review?
    - Conceptual?
    - Order of Magnitude
    - Pre-Feasability
    - Etc
- Baseline PHA?
- Revalidation PHA?
- A post incident review?
- Etc.

The methodology of review should also be specified, is it a HAZOP, HAZID or other methodology. The company Subject Mater Expert should be involved in the decision.

2.1.2 Scope of the review

It is important to specify what is in the review and what is excluded. Where there are multiple linked Process Hazard areas, it is important to specify which equipment are part of which PHA, as interconnecting piping are often missed.

2.1.3 Objectives of the review

2.1.3.1 Primary review objectives

In all cases, the objectives of the review should state the processes reviewed meet the company tolerances once all recommendations have been put into place. For project reviews, the objective should be appropriate to the advancement of the project. The objective should be different for a project in Conceptual Engineering versus a project in Detailed Engineering.

2.1.3.2 Secondary review objectives

There may be secondary review objectives such as updating Emergency Response Planning based on PHA scenarios where required. Update P&ID's and PFD's based on PHA review mark-ups.

The objectives portion of the Charter an appropriate place to list the deliverables, for external contractors, these should also be listed in the contract, the Charter should include dates and names:

- Executive summary
- Scope of the review
- List of information / documents used during the review
- Assumptions used
- Short description of the methodology used during the review
- List of attendees with expertise
- List of recommendations
- Detailed review sheets
- Etc

2.1.4 Preparation of the review

Reports

Previous PHA reports should be available during planning for the PHA. For projects, the PHA should build on previous stages, Detailed Engineering building on Pre-Feasibility. For existing plants, Revalidations should build on Baseline, Baseline should build on Project PHA etc.

Incidents

For existing systems, all incidents from the equipment under review should be listed in the Charter. For revalidations, all incidents since previous PHA study, being the Baseline, or a previous Revalidation. All incidents on similar company systems should also be included whether they are from the site off interest or other company sites. Available incidents occurring in similar equipment from industry should also be included.

For project PHA's, a list of incidents occurring in similar facilities should be included. For research projects, it is often said that they are a "one of". However, these processes normally use common unit operations, sometimes in a novel way, or with novel chemicals, there are typically incidents.

Incidents are used to maintain a level of Chronic Unease to fight the tendencies that many have that the PHA scenarios are not credible. If it has happened on site or in the industry, why can it not happen here?

2.1.5 Review team

The review team must have all the required experience around the table to conduct the review. For smaller sites the same person might have multiple expertise.

The experience type that we are looking for are the following:

- Facilitator (approved)
- Scribe (understanding the equipment within the scope)
- Process (understanding the design and theory of operations)
- Operations (experience operating the equipment within the scope)
- Mechanical (understand repairs, corrosion, failure mechanisms)
- Instrumentation (understand the details around how the equipment within the scope is controlled)
- Health Safety and Environment (HSE) (experienced with the impacts of incidents)
- Other specialists as required, such as the technology provider SME
- All participants should come prepared for the review with the appropriate **knowledge** and **documentation**.

Keeping in mind that the optimal size of the review team is six to eight participants, for large projects where there are site, engineering and technology provider representatives, this number can easily reach 15 to 18 plus participants. Having these many people extends the review length and complicates the review dynamics.

For sites starting on the PSM journey, it is recommended that they get help setting up their PSM processes. A good practice for sites external facilitator, is to have them approved by a SME since there is a fair variability in PHA facilitators. This is suggested to insure the reviews follows all RAGAGEPS.

2.1.6 Documentation

Depending on whether the review is a conceptual HAZID of a detailed design PHA, the required Process Safety Information. Generally, a typical list of required PSI is noted below: (responsible discipline)

- Process Flow Diagram (P)
- Piping and Instrumentation Diagram (P)
- Area Electrical Classification (if required) (P)
- Flame, Flammable, and toxic detection layout (P)

- Functional Description of the controls, including any Safety Instrumented Systems (SIS) (I)
- Heat and Material Balance (P)
- Design Basis (P – M)
- Mechanical equipment drawing, including pump curves for centrifugal pumps (M)
- Material Safety Data sheets (P)
- Chemical interaction matrix (P)
- Operating and maintenance manuals (P – M)

P – Process
M – Mechanical
I – Instrumentation
H – Health Safety and Environment

The PHA organizer should gather all the documentation prior to the review.

All these documents need to be up to dates. For these reasons, groundwork may be required to evaluate the quality of this information.

2.1.7 Dates of the Review

The review dates should be included in the charter, specifying dates, times (start and end of each day), including time zones when required, and a preliminary agenda.

When developing the Charter, the site will need to work with the facilitator (either internal or external) to understand the length of the review process, which will only be available once the process has been divided into nodes. The definition of a node is a group of equipment that has the same function. The process needs to be divided into nodes in order to apply the review methodology. The node size will depend on PHA methodology.

When choosing the dates of the review, it is important to also take into consideration the physical aspects of the review. During Covid, we needed to adjust to conducting these reviews remote (by Teams, Webex, Skype or other methods). There are many benefits to conducting PHA's in person versus remote, so where possible, the in-person review should be privileged.

The ideal conference room is large enough to accommodate the review team while not being to large so that communication is easy. It is quiet enough so not as to not be negatively impacted by other conference room noise, construction noise etc.

Having the ability to place the reference drawing on the wall improves the review dynamics by permitting the team to get up and aggregate around the drawings. Site access should also be considered. Being close enough to be able to conduct a site tour but being far enough as to not loose participant during the review.

2.1.8 Methodology

The review methodology should reference the company standard or a recognized standard such as the Guidelines for Hazard Evaluation Procedures, 3rd Edition from Wiley (CCPS), or HAZOP: Guide to Best Practice, 3rd Edition by Elsevier (IChemE), or equivalent.

The different receptors of interest for the review should also be stated in the Charter, common ones are:

- Health and Safety
- Environment
- Company assets
- Reputation
- Community
- Conformance

Some companies only evaluate Health and Safety and Environment, and do not evaluate the other receptors during the PHA.

It should also be mentioned in the Charter which operational states will be covered:

- Normal operation
- Start-up and shut-down
- Emergency shutdown
- Maintenance

2.1.9 Risk tolerance

It is up to the company carrying out the PHA to determine their risk tolerance. If they do not have a risk matrix, it is recommended that they get support in establishing their corporate tolerance, as this will drive the costs of the recommendations. Many smaller companies do not have a corporate risk matrix.

3 Runing the review

3.1 Introductions and presentation

At the beginning of the review, an introduction will be given providing attendees with the following information:

- Scope of the review
- Agenda
- Attendees
- Brief introduction to the methodology used for the review (training for unexperienced participants)
- Outcomes of the review: risk profile, actions, process safety scenarios
- Principles of recommendations / actions proposed
- Golden rules during the review

3.2 Execution of the review

When planning the review, team dynamics impact review duration. Teams with little or no PHA experience take up to twice the time seasoned teams take to cover the same number of nodes, especially if nodes are complicated, this should be taken into account during the planning.

The process engineer or equivalent person knowledgeable of the equipment being reviewed should describe how the process works to align attendees to a common understanding. A good practice is to explain the overall process, then to explain each individual node focussing on its operation modes.

Once the node has been described, the actual review can start with the documentation of the methodology.

A best practice is to document three levels of risk;

- **Inherent** risk, this is the risk without safeguards, it is often the initiating event frequency. This risk (severity and frequency) is used to understand the number of independent safeguards required to meet tolerance
- **Current** risk, this is the risk with all safeguards applied. Care should be taken here to document independent safeguards which contribute to risk reduction and non independent safeguards which do not contribute to risk reduction, and why.
- **Future** or proposed risk once all recommendations have been implemented. This tolerance should meet the company's tolerance.

For longer reviews, it is recommended to go over the recommendations prior to presenting them to people outside the PHA Team. Attributes of a good recommendations are:

- Should be in a standalone format, referring to the scenario of concern i.e. toxic release caused by overpressure of…
- Should be directive, using words change, add, modify to prevent…
    - Study, Evaluate, Consider or similar words should not be used.

The details of the execution of the study will depend on the companies PHA procedure. It is not the intent of this paper to describe how to execute the PHA, but rather how to prepare for it, and how to wrap it up.

4 Post review

4.1 Goals and measures

- PHA Close-out meeting.

  It is a good practice to hold a close-out meeting on the last day of the review with the PHA sponsor and PHA team. In this meeting, the major findings (recommendations) of the review should be presented.

- PHA draft report for comments within one to two weeks, depending on PHA complexity.

  The time for issuance of the draft PHA report for comments should be specified in the Charter, as the longer it takes to issue the report, the more likely the site will lose some PHA focus, as the team moves on to other priorities.

- The final report issued within four to six weeks of PHA's completion.

  Allow two weeks to receive site comments once preliminary report is issued. Incorporate any comments appropriate comments received from the site and issue final.

- PHA's completion / quality checklist issued with final report.

  There remain many activities coming from the PHA. A completion / quality checklist is a good way to document what is completed or not.

    - Are all the deliverables detailed in the charter included in the report
    - Marked-up PSI, mainly P&IDs sent to drafting following the site update procedure

5 Conclusions

This paper documents current industry best practices for setting up a PHA for success. All the activities described in the paper require time (money). Some companies at the start of their PSM journey may not have value for all activities. The more the PHA process incorporates, the higher probability of success.

SPRING24
+20TH GCPS
A Joint AIChE and CCPS Meeting

National Grid's Management of Change (MOC) and Action Item Automation

Enita Okodiko
National Grid
170 Data Drive, Waltham, MA
Enita.Okodiko@NationalGrid.com

Xavier Strittmatter
National Grid
170 Data Drive, Waltham, MA
Xavier.Strittmatter@NationalGrid.com

Abigail Calistra
National Grid
170 Data Drive, Waltham, MA
Abigail.Calistra@NationalGrid.com

Prepared for Presentation at
American Institute of Chemical Engineers
2024 Spring Meeting and 20th Global Congress on Process Safety
New Orleans, LA
March 24-28, 2024

National Grid's Management of Change (MOC) and Action Item Automation

Enita Okodiko
National Grid
170 Data Drive, Waltham, MA
Enita.Okodiko@NationalGrid.com

Xavier Strittmatter
National Grid

Abigail Calistra
National Grid

Keywords: Management of Change, Action Item

Abstract

National Grid aims to be a leader in process safety management in the industry by implementing risk control standards for its natural gas storage, transmission, and distribution assets. The company's Process Safety Management System framework mandates that all new and modified assets be assessed using the Management of Change database and Action Item Log to ensure protection of employees and members of the public, environmental conservation, and regulatory compliance. This paper describes the evolution of National Grid's management of change and action item management systems, highlighting the challenges posed by the company's previous manual system and the need for automation. It provides an overview of the requirements for the new software solution, which focuses on creating dynamic management of change forms, automating workflows, and providing data reporting capabilities. The paper also discusses the benefits of the new system, including improved efficiency, streamlined processes, and a strengthened process safety culture. Overall, this paper offers insights into the process of automating management of change and action item management systems and emphasizes the importance of effective process safety management in ensuring the integrity and safety of National Grid's gas business unit.

1 Introduction

National Grid is committed to pioneering innovation in energy management and preserving our global environment for future generations. Our primary objective is to achieve this through a dedicated and proficient workforce, supported by processes that effectively manage and control our operations and assets. To realize this commitment, we rely on well-defined procedures and a robust safety management system.

In 2015, National Grid introduced a comprehensive Process Safety Management System (PSMS) across the entire organization. This system further reinforces our ongoing dedication to Process Safety and encompasses a Group Framework, a Process Safety Approach for Managing Major Asset Hazards, and twelve Risk Control Standards. The group wide PSMS mandates that all business units operating major hazard assets, particularly the gas business unit, establish and uphold systems and practices in compliance with these expectations, including the crucial aspect of management of change (MOC) and action item management.

A resilient safety management system is imperative to prevent and mitigate major process safety incidents. Effective management of change and action item management is paramount in ensuring the integrity and safety of our gas business unit. Within the PSMS framework, the management of change becomes even more critical, playing a pivotal role in incident prevention, protection of employees and the public, environmental conservation, and regulatory compliance.

This white paper aims to trace the evolution of the management of change and action item management within National Grid's PSMS for the Gas Business unit. It will elucidate key concepts, highlight pain points, and provide insights into the implementation of best practices. By embracing these practices, National Grid's gas business unit can systematically identify, evaluate, and control the risks associated with any proposed changes in their processes.

2 Background

The framework of the company's PSMS requires assessment of all new assets and modifications to existing assets prior to operation using its MOC process. National Grid's MOC process is designed to:

- Ensure all changes are evaluated, documented, and approved when required before implementation of the change.
- Ensure that the potential process safety implications, such as societal, individual, and environmental risk, of the change are reviewed.
- Identify the potential need to update Process Safety Information (PSI) to reflect the change (e.g., P&IDs, employee training, etc.).
- Ensure that all changes are reviewed and authorized by competent personnel.

The MOC process identifies when additional risk assessment tools, such as Facility Siting Studies and Process Hazard Analyses (PHAs) are required. It also ensures that the proper personnel are notified of a proposed change and are given the chance to review the effect a change may have on the operation of the system. This process includes assigning responsibility for follow-up actions to the appropriate individuals within the business. For example, a risk identified in a PHA may require a design change to incorporate a relief valve to avoid over-pressurization of a section of steel pipe. This action would be assigned and tracked via the Action Item Log. The action item management process is defined in National Grid's process safety procedures and covers many action item sources besides just the MOC process. Once an individual completes an action, they provide documentation of the completed action item, and must receive approval from their manager and the person responsible for managing the project (typically a project manager or asset engineer).

Historically, National Grid utilized a SharePoint 2010 site to manage its MOC process. The system relied on manual notification of a submitted change to the MOC coordinator, approvers, and key stakeholders, which impeded the efficiency of the approval process. Similarly, the action item process utilized an Access database, which required manual due date tracking, manual notification of when an action was completed, and sign-off from the approvers via email or paper form. The Access database was only accessible by certain personnel, so visibility to upcoming action items was limited. If the manual process of communicating and coordinating actions broke down, action item owners would not even know they had been assigned actions due to this lack of transparency. National Grid's focus on process safety has grown each year leading to a 550% increase in MOCs and action items in the system between 2017 and 2022. This exacerbated the challenges faced by the gas business's process safety team as each member of the team were required to spend 28 hours per week to manage the MOC and Action Item Libraries. The challenges of a manual process resulted in a continually increasing backlog of overdue actions, reaching nearly 300 overdue action items by January 2021. Additional resources were needed by the gas process safety team to reduce the overdue action items. However, the gas business still struggled to stay on top of the backlog as personnel changes had a significant impact on the ability to maintain and update the MOC and Action Item Libraries in a timely manner. There was a clear need for automation of this system as it continued to become a more inefficient and unsustainable use of the company's process safety resources.

The need for an updated system was further exacerbated by several other factors. SharePoint 2010 support was discontinued by Microsoft in April 2021, which presented security and operability risks with continuing to use that platform. The employee who had originally created and managed the old MOC SharePoint had left the company, and some of their permissions were not transferred correctly, effectively preventing the update of certain features. National Grid's MOC and Action Item management policies had not been updated since the SharePoint 2010 site was originally published. Some of the unchangeable features of the old site no longer met the requirements of the updated corporate policies. The old site utilized a static form for MOCs, meaning the questions and sections were the same for all asset classes and change types. This led to user input errors. Users would be unsure which sections applied to their change type, leading them either to fail to fill out critical information about the process safety implications of an asset or to fill out unnecessary information, wasting valuable resources.

All these shortcomings with the two existing platforms were considered when developing a strategy to improve the MOC and Action Item libraries. Lessons-learned were incorporated into the design of the new platform. Notably, the following were focus areas for the new platform:

- Creation of dynamic forms that respond to user input so only relevant questions are asked and correct information is captured.
- Allowing user roles to be easily updated if someone leaves the company or changes job duties.
- Automation of workflows for processes like approvals and due date extension requests.
- Sending automatic notifications via email and in-app.
- Allowing for reporting capabilities and data visualization via PowerBI.

3 Steps for Automation

The first challenge in project development was to obtain business approval from the company's leadership team. To develop this business case, the team researched various software alternatives to the existing system; engaged with key stakeholders and users of the system to determine what changes would add the most value to the business; and defined the proposed upgrades in the scope of work documentation. Next, more detailed documentation for the implementation of the project was developed: including the mapping of the original process workflow, an automation logic map, and test cases for the new platform. Mapping out processes using flow charts and lists was key to fully understanding how the MOC process should perform and what features were required in a new software solution. Figure 1 below is an example of a flow chart developed for this purpose showing how an MOC would progress through different statuses.

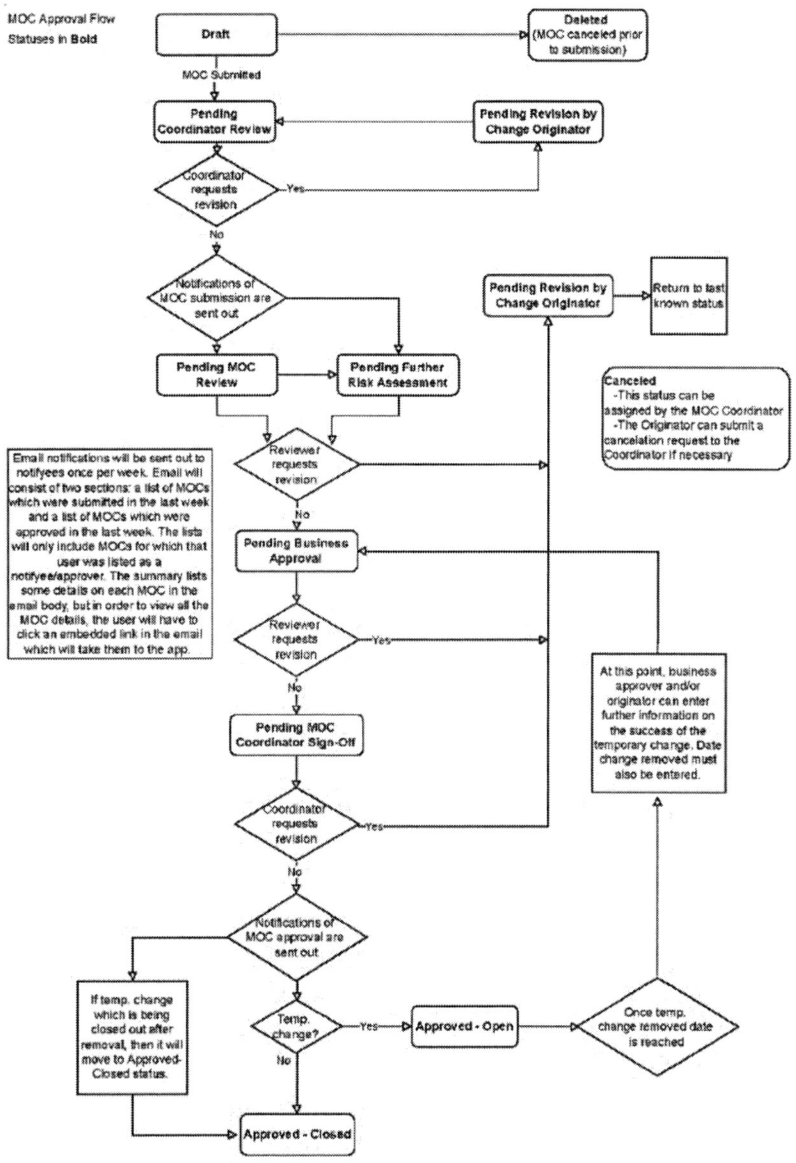

Figure 1 – MOC Status Flow Chart

After review of the required features and workflows for the new system, Microsoft's Power Platform was identified as a suitable host. Power Platform was chosen for several reasons:

- National Grid already had the software licensing, making this the most affordable option.
- National Grid had in-house expertise in development and support for this software.
- Microsoft Power Apps applications are customizable, allowing the software to be built for a specific purpose.
- Power Apps is integrated with the rest of Microsoft's product line, allowing easy integration with Microsoft Teams, Outlook, and SharePoint to store and manage data and access employee data.

A software development team and project manager were provided by National Grid IT to assist in the development of the new platform. Additionally, some engineers, project managers, and supervisors were user testers. Throughout the process, documentation of technical requirements was vital to ensure project success. The software development professionals working on the actual product had little-to-no understanding of what a process safety management system entailed, nor any of the specifics of an MOC process or action item library. The project leads (a member of the design engineering team and a member of the process safety team) had to painstakingly outline every minute detail of the process for the development team to holistically understand how the final product should perform. Inevitably, details were missed, and requirements were misunderstood, so having well-documented project requirements in writing was key to resolving these issues. Acceptance testing was performed by the primary project leads and user testers. The test cases needed to incorporate both "happy path" and "unhappy path" scenarios. A happy path is when a user fills out all the checklists and forms as expected and without input errors. An unhappy path is when a user does not fill out all required information or tries to use the software in an unintended way. Some examples of test cases are shown in Table 1 below.

#	Test Case	Testing Steps
1	"Happy Path" - RIK - PRE - DNY	1. Create MOC 2. Select Temporary Change - select a due date in the near future 3. Select Replacement-In-Kind a. Answer "No" to questions 1-8 on RIK checklist 4. Add others in the meeting as Notifyees 5. Add W.W. as a Business Approver 6. Submit and Test Workflow - Approvers should approve at every step
2	"Unhappy path" - Minor Change - GTE - MA	1. Create MOC 2. Select Emergency Change 3. Save as draft multiple times throughout the process without having completed the form 4. Do Not Select Replacement-In-Kind a. Answer "No" to questions 1, 2, 4, 6, 9-11, 20, & 21 on the SHE Review Checklist 5. Add others in the meeting as Notifyees 6. Add J.B. as Business Approver 7. Submit and Test Workflow - At every step, approvers should send back for revision

Table 1 – User Acceptance Test Cases

Numerous software bugs were caught throughout the testing process, mostly during unhappy path testing scenarios. Most of these bugs could be attributed to a lack of thorough documentation of requirements. Many details that seemed insignificant or intuitive to a design engineer or a process safety professional were missed in the documentation, but the lack of detail would lead to poor assumptions by the software development team. Resolving some of these issues required compromise between the software team and the project leads due to budget and project timeline constraints. After testing was completed and bugs addressed, the migration of data to this new platform began in October 2022. The final platform was launched in November 2022 after over a year and a half of work from proposal to implementation.

4 Process Improvements

The updated system focuses on creating an intuitive workflow that sends automatic email notifications to all necessary parties at key stages of the process. Operations personnel, engineering managers, asset engineers, the gas control operators, and the gas planning teams receive notifications of a proposed change in the new system, which allows them to provide feedback on the engineering design and raise any safety concerns with the proposed change. The email notifications also keep the workflow on track by alerting managers and process safety personnel when their approval is required and by alerting the change originator (typically the design engineer) when action is required from them, such as revising a design document. The old systems did not have any automated notifications, and oftentimes MOCs would sit in limbo for months with no action because the responsible party was simply uninformed. Picture 1 below shows an example email notification generated by the new system.

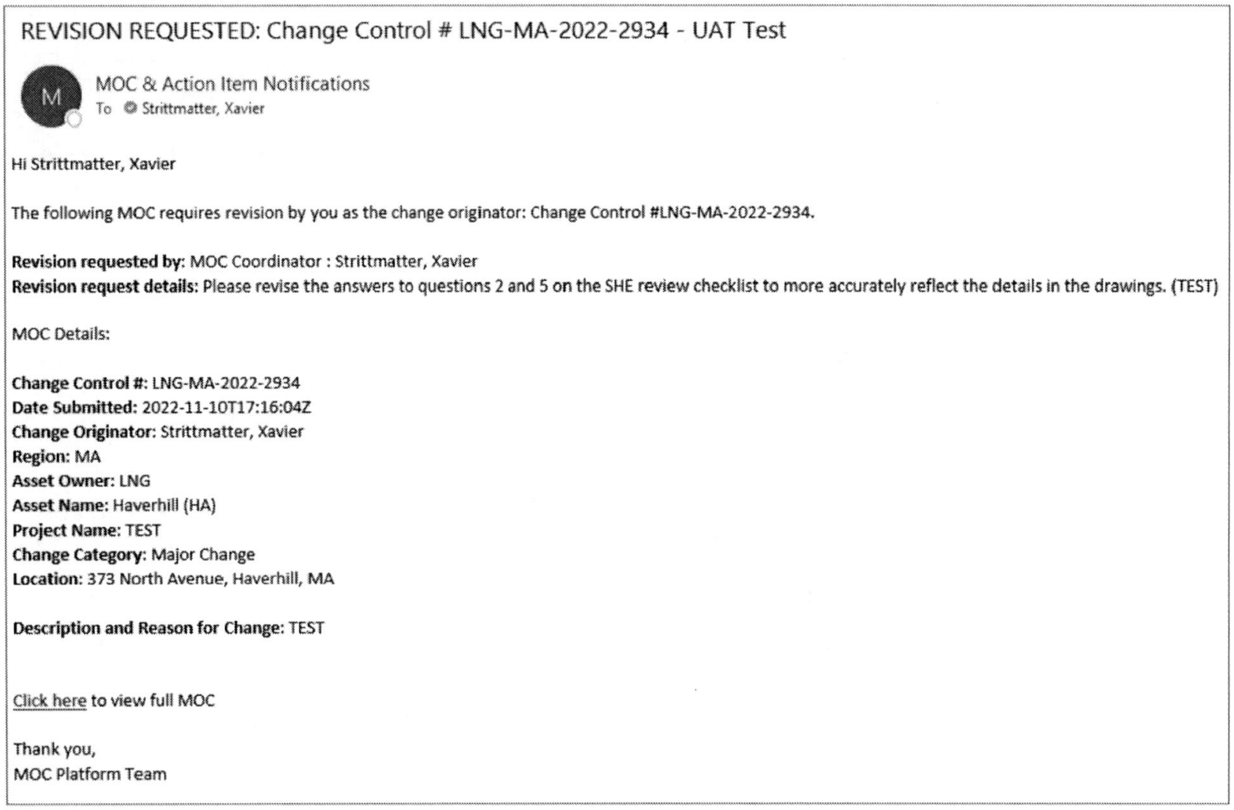

REVISION REQUESTED: Change Control # LNG-MA-2022-2934 - UAT Test

MOC & Action Item Notifications
To Strittmatter, Xavier

Hi Strittmatter, Xavier

The following MOC requires revision by you as the change originator: Change Control #LNG-MA-2022-2934.

Revision requested by: MOC Coordinator : Strittmatter, Xavier
Revision request details: Please revise the answers to questions 2 and 5 on the SHE review checklist to more accurately reflect the details in the drawings. (TEST)

MOC Details:

Change Control #: LNG-MA-2022-2934
Date Submitted: 2022-11-10T17:16:04Z
Change Originator: Strittmatter, Xavier
Region: MA
Asset Owner: LNG
Asset Name: Haverhill (HA)
Project Name: TEST
Change Category: Major Change
Location: 373 North Avenue, Haverhill, MA

Description and Reason for Change: TEST

Click here to view full MOC

Thank you,
MOC Platform Team

Picture 1 – Email Notification Example

The new MOC form is configured for each specific asset type, providing automatic prompts for the change originator based on their responses to previous questions. By providing a more comprehensive explanation of what is required for each field, change originators get a better understanding of why MOCs are an essential part of the PSMS, how the process relates directly to their asset class, and how it enhances their project. Dropdown menus featured in the MOC form reduce user entry errors which decreases the time spent by MOC coordinators checking submitted forms for grammatical errors and reduce time spent by change originators typing. Picture 2 below shows an example of a dropdown menu. The ability to track MOC status has also greatly improved with the addition of library search and filtering features based on various criteria like region, asset owner, workorder, and MOC status. The old library was not searchable based on any of these criteria. Picture 3 shows the new MOC library search screen.

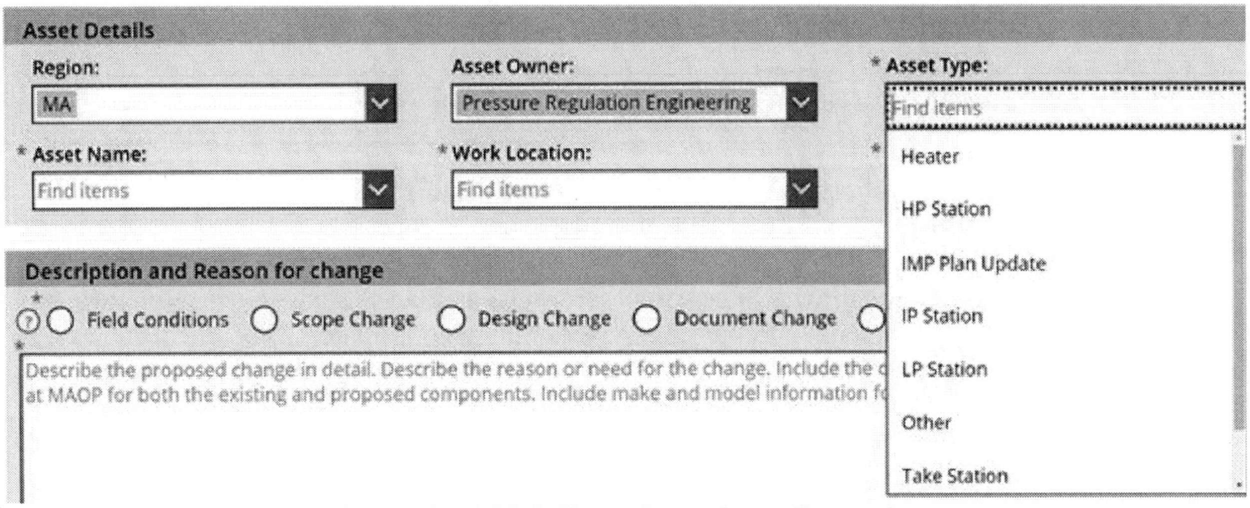

Picture 2 – MOC Form Dropdown Example

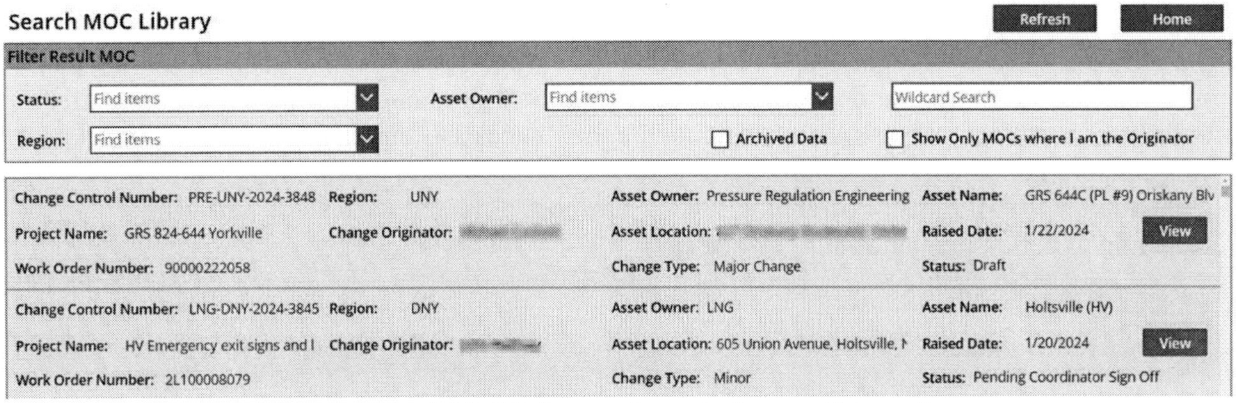

Picture 3 – MOC Library Search Screen Example

Another big improvement in the new system was the addition of tooltips, question mark icons that could be hovered over to read more information about a question on a form or checklist within the MOC process. Tooltips were also used to give the definitions of process safety terminology used within National Grid internal policies and procedures so users could accurately understand and fill-out information without referencing additional documents. The old system suffered from a lack of user knowledge as many did not have to fill out MOCs regularly. This led to input errors, insufficient data entry, or incorrect data entry. The addition of tooltips has lowered these user errors and given users a better understanding of the MOC process.

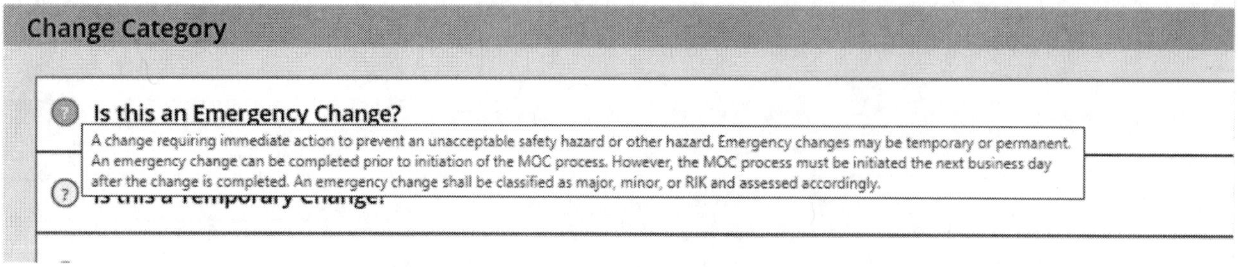

Picture 4 – Example of Tooltip Hint Text

Action items generated by MOC forms are now automatically populated into the Action Item Log. Action items generated through other processes, such as PHAs, Management Reviews, and LOPAs, can be manually entered into the Log by action item coordinators. The identified owner of an action item is automatically notified by email and given the opportunity to confirm their ownership or indicate who else should be responsible. Action item owners receive automated email reminders of upcoming due dates on a defined timeline: 1 year out, 6 months out, monthly for the last three months, then weekly for the last month. The notifications also include an escalation mechanism, notifying the action item owner's supervisor that their employee has an action item coming due very soon or overdue. Owners of action items can close their items or request due date extension within the system. The added workflow capabilities for the action item process have resulted in a near elimination in overdue action items. Furthermore, the new platform allows for adaptability. Any changes that are required due to corporate process safety updates or specific asset teams' desire to update their checklists can be implemented by MOC coordinators or National Grid's IT support. Additionally, the new Power Apps platform is compatible with Microsoft PowerBI, allowing for more efficient, consistent, and visible reporting of action item and MOC statuses.

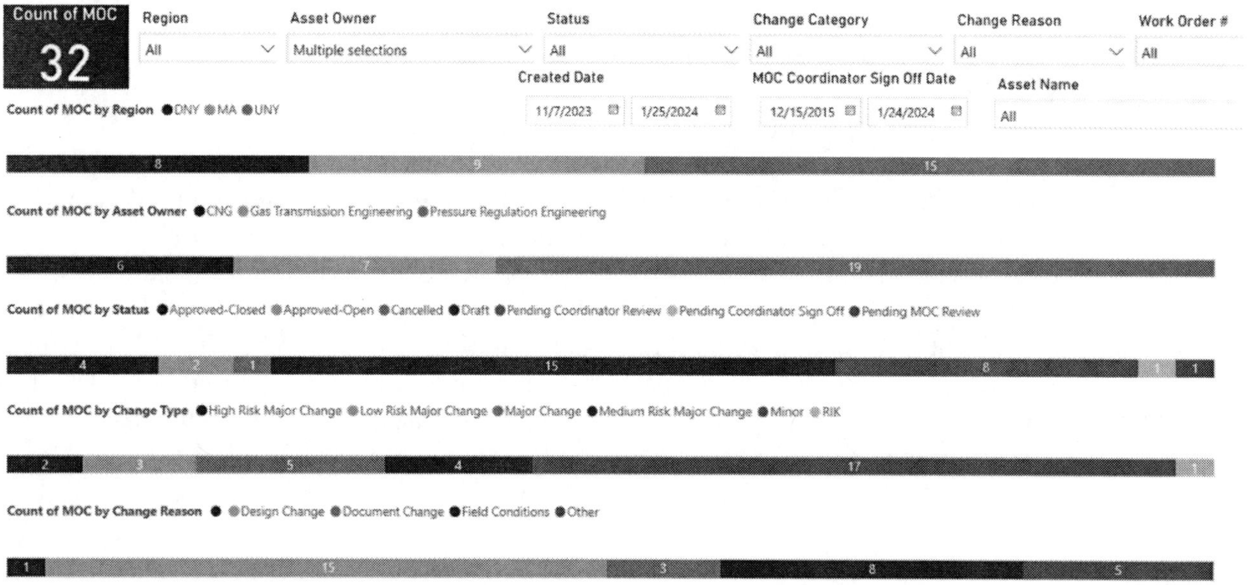

Picture 5 – PowerBI Dashboard for MOC Metrics

By streamlining the process, the company reduced the amount of time spent managing the MOC and Action Item Libraries. This has allowed the gas business process safety team to spend more time on risk-based activities such as Process Hazard Analyses (PHAs), Layer of Protection Analyses (LOPAs), Quantitative Risk Assessments, Facility Siting Studies, and high-level communication of risks and safety focus areas.

5 Effects on Process Safety Culture

With the integration of the Management of Change (MOC) and Action Item automation platform, National Grid Gas Business marks an important milestone in the ongoing development of its process safety culture. The company initially set a conservative target of two MOCs per month following the implementation of the process safety management system in 2017 but has since exceeded this goal by observing a significant increase to over thirty MOCs per month. Not only is this upswing a result of the intuitive MOC platform, but also of the company's steadfast commitment to meeting regulatory requirements. With the seamless integration of the MOC platform into daily operations, the organization has developed a more agile and responsive approach to managing change.

There can be no overstatement of the importance of the automated notification system. Aside from streamlining the MOC process, it has expanded engagement to include non-traditional groups. In particular, field personnel, who were previously peripheral to the MOC process, now actively participate in the process by entering MOCs independently into the system. As a result of this democratization, the process of managing change has been fundamentally altered within an organization. As a replacement for the former manual notification process, prone to bottlenecks and inefficiencies, automatic notifications ensure timely and comprehensive communication. Through this shift, stakeholders are now able to participate in decision-making during times of change in a collaborative and inclusive environment.

Regarding the Action Item aspect, the automation feature has emerged as an important catalyst for organizational agility. Assigning responsibility and empowering individuals to oversee item closures have resulted in a paradigm shift. There are now groups with limited exposure to the intricacies of the company's process safety management system actively participating in the implementation of one or more of the company's risk control standards. In addition to aligning with broader organizational goals, this contributes significantly to the maturation of the overall process safety culture within the company.

Combined with sophisticated tools such as PowerBI, the integration of the MOC and Action Item Automation platform has ushered in a new era of transparency. The status of MOCs and action items is now more visible to all groups, including frontline workers and leadership. An important aspect of this transparency is the ability to track overdue items, upcoming changes, approved modifications, and more, which plays a pivotal role in managing the company's process safety risks. Enhanced transparency facilitates informed decision-making, enabling the organization to navigate the complex landscape of process safety with confidence.

As a result, the transformative impact of the MOC and Action Item Automation platform, coupled with advanced tools, extends beyond simple procedural improvements. It has proven to be a cornerstone of the organization's continuous evolution of its process safety culture, resulting in the achievement of key targets and the effective implementation of the company's strategic plan. It is National Grid Gas Business' commitment to safety, efficiency, and excellence in the dynamic landscape of process safety management that has led the company to continue to embrace technological innovations.

6 Conclusion

In summation, automation of the MOC and action item processes was essential for the continued growth of National Grid's implemented PSMS. It allowed for gains in efficiency as well as bolstering the company's overall process safety culture. These processes are crucial elements for the gas business unit to prevent and mitigate incidents that could be harmful to employees, members of the public, and the environment by assessing and managing the risks related to implementing new assets and modifying existing ones.

The existing SharePoint 2010 system relied on manual notification and a single MOC form for all asset types which slowed down the process leading to a backlog of overdue action items and the mismanagement of other process safety project elements. These issues shaped the requirements for determining the new software solution. The search for the new platform focused on creating dynamic MOC forms, making it easier to transition personnel roles, automating workflows with automatic email notifications, and providing data reporting capabilities. Microsoft's Power Platform was identified as a suitable host due to affordability, in-house expertise, and customization options. After the creation and beta testing of the platform, it was implemented in November 2022 after a year and a half of work.

The automated MOC and action item platform has allowed for a more agile and responsive approach to managing change. This has resulted in a paradigm shift within the business, empowering individuals to oversee item closures and contributing significantly to the maturation of the overall process safety culture. The integration of this automated platform has ushered in a new era of transparency, which has enabled the organization to navigate the complex landscape of process safety with confidence.

7 References

1. OSHA. 1910.119 - Process safety management of highly hazardous chemicals. | Occupational Safety and Health Administration. Osha.gov. Published 2012. https://www.osha.gov/laws-regs/regulations/standardnumber/1910/1910.119
2. American Institute of Chemical Engineers. Center For Chemical Process Safety. Guidelines for Managing Process Safety Risks during Organizational Change. Center For Chemical Process Safety, Wiley, A John Wiley & Sons, Inc., Publication; 2013.
3. American Institute of Chemical Engineers. Center For Chemical Process Safety. Guidelines for the Management of Change for Process Safety. John Wiley & Sons; 2008.
4. Occupational Safety and Health Administration. Process safety management of highly hazardous chemicals. Published February 8, 2013. https://www.osha.gov/laws-regs/regulations/standardnumber/1910/1910.119
5. National Fire Protection Association. NFPA 652: Standard on the Fundamentals of Combustible Dust. Published 2019. https://www.nfpa.org/codes-and-standards/6/5/2/652

111f Integrated Approach for HAZOP, LOPA and Alarm Rationalisation

Khama Matiti

The oil and gas industry is constantly evolving and growing in complexity, with inherent temperature, pressure, and hazardous risks. To maintain safety and stakeholder confidence, a systematic and auditable approach to plant design and operation is essential. Operators are pursuing strategies to enhance profitability and efficiency, particularly in the LNG sector, which is in high demand due to its environmental benefits.

Process Hazard Analysis (PHA) is widely used to identify potential hazards in engineered designs. Various PHA analyses are typically performed during different phases of facility development. Integration of safety analyses like HAZOP, LOPA, and Alarm Rationalisation is essential, but often they are conducted in isolation by different disciplines, leading to inconsistencies and contradictory recommendations.

To address these issues and ensure safe LNG plant design, a formal, systematic, and integrated system for HAZOP, LOPA, and Alarm Rationalisation was developed. This approach minimizes inefficiencies, streamlines analysis, and improves the quality of findings. The paper discusses the application of this methodology in an LNG project, adhering to international standards. The technique will demonstrate how to ensure an integrated approach to HAZOP, LOPA and Alarm Rationalisation whilst following the international standards namely IEC 61882 (Cenelec, 2016) , IEC 61511 (International Electrotechnical Commission,2016) and IEC 62682 (International Electrotechnical Commission,2014).

133b Improving Process Safety Culture at a Major Natural Gas Pipeline Company

Tekin Kunt, Seckin Gökçe, Sinan Gunesli, Dilek Akbey

This presentation highlights the successful implementation of a Process Safety Management (PSM) system improvement initiative at the Trans Anatolian Natural Gas Pipeline (TANAP). TANAP is part of the Southern Gas Corridor which is extending from Caspian Sea to Italy. TANAP transports Natural Gas from the Shah Deniz gas field of Azerbaijan, and possibly from other neighboring countries to Turkey and through TAP Pipeline to Europe. TANAP recognizes the utmost importance of maintaining a robust PSM framework to ensure safe and reliable operations.

The PSM program improvement initiative aimed to enhance various aspects of the PSM system through rigorous assessment and analysis for potential areas for improvement. These areas spanned a wide range of elements, including management of change, process safety culture, measurement and metrics enhancements, process knowledge management and process safety competence assessments.

This paper will focus on the activities conducted to improve Process Safety Culture in the Company. Key to the success of this endeavor was the strong support and commitment from the TANAP executive management. Their active involvement, leadership, and allocation of necessary resources have fostered a culture of safety, facilitating the integration of enhanced PSM practices into the organization. This collaborative approach has resulted in notable advancements in the process safety performance, operational efficiency, and risk reduction within TANAP.

This presentation presents a comprehensive overview of the efforts taken by TANAP in improving the PSM system. It highlights the importance of a strong management commitment to process safety and provides valuable insights into the strategies, challenges, and outcomes of improving process safety culture in an organization. The lessons learned and best practices derived from this initiative can serve as a reference for other organizations seeking to enhance their process safety culture.

133c Process Safety Site Assessment Program (PSSAP(r)): Repeat Assessments Drive Improvements in PSM

Colin Frazier, Ryan Meng

The Process Site Assessment Program (PSSAP®) is a not-for-profit industry assessment program managed by API, focused on sites that refine or process hydrocarbons, that conducts assessments around the world to help mature process safety programs for over a decade. To date, API has completed over 175 assessments at sites all over the world, inclusive of refineries, petrochemical plants, liquefied natural gas facilities, and tank farms.

During a PSSAP assessment, a team of independent, third-party process safety experts (with an average of 40+ years of experience, each) spend about a week at a site, working with employees and leadership — across shifts — to fully evaluate the health of the existing process safety programs. They use our industry-created good practice-based protocols to systematically provide feedback to a site on a range of process safety topics:

- Process Safety Leadership
- Operating Practices
- Mechanical Integrity (focused on fixed equipment)
- Safe Work Practices
- Management of Change (MOC)
- Process Hazards Analysis (PHA)
- Facility Siting
- Product Storage & Transfer
- Incident Learning
- Hydrofluoric Acid (HF) Alkylation/API RP-751
- Safeguards (available if completing PHA protocol)

The PSSAP protocol topics listed include over 750 individual requirements. They also reference over 100 standards, inclusive of API and non-API standards, as appropriate. The assessors also score each of these 750+ requirements for how well they are implemented at a site. As a result, we have a consistently applied methodology for our benchmarking efforts. The benchmarking database houses the scores of site performance in a blinded and anonymous way, allowing API to analyze the industry performance data and identify trends. By participating in a PSSAP Assessment, a site gains access to this database and compares their results to the results of their peers, blinded and anonymously (i.e., no site can discern the performance of another site). In this presentation, a discussion about the program, the benchmarking, and examples of previous benchmarking will be shown.

API continues to grow this program and expand, having conducted 23 assessments in 2022 and 17 assessments in 2023. We will discuss the assessment trending by types of operators as well as in countries of entry - API has currently conducted assessments in 12 countries around the world.

In the past few years, several sites have been returning to API to have repeated assessments conducted at their sites, presenting a unique opportunity to benchmark their progress toward implementation of the 600+ PSSAP General Assessment requirements (7 of the 10 protocols make up a General Assessment). To date, eleven (11) sites have conducted repeated General Assessments. Of those, ten (10) showed an increase in performance on their second assessment (overall score). On average, scoring on the Mechanical Integrity protocol increase by 6.9% - an increase indicative of large changes and performance improvements in managing these risks at the sites assessed. API is encouraged by these results on repeat assessments.

A few key factors are critical for success in continuously improving process safety programs: 1) a site must have support from senior leadership that the assessment's results should be actioned; 2) a site's culture should embrace continuous improvement and treat the assessments as learning opportunities, not like they would regulatory audits; 3) turnover of leadership and subject matter expertise in this area can have a large impact in the success of the continuous improvement efforts. These will be expanded in more detail as part of the presentation.

API continues to market these programs to all industries that process hazardous materials, especially hydrocarbons. We have created an industry-led offering that continues to drive process safety performance in the right direction, and look forward to continuing to expand into other subsegments of the industry.

133e Mastering the Art of Decision Making in Challenging Environments

Chris Scott, Gareth Black, Bethany Elliott

In a rapidly evolving industry where enhancing safety and performance is key, the need for effective decision making is a critical skill for individuals and organizations alike. This interactive workshop aims to equip participants with the tools and strategies necessary to make informed and confident decisions in challenging environments. Through a combination of theoretical insights, real-world case studies, and hands-on exercises, attendees will learn how to navigate complex choices, manage uncertainty, and optimize outcomes. Whether you're simply curious to learn more about the challenges of high stress environments or you’re a leader responsible for guiding your team to enact Responsible Care initiatives, this workshop will provide valuable insights and practical techniques to improve your decision-making abilities. Join us to unlock the secrets behind making sound decisions that drive success and innovation in environmental, health, safety, and security (EHS&S) performance.

168a The Invisible Hazard: Exploring the Impact of Stress and Personal Resilience on Safety in the Chemical Industry

Bethany Elliott

In recent years, stress has been a buzzword across industries. Are people more stressed now than they were before, or are we finally paying attention to the risks associated with stressed out employees? Whatever the answer, the increased attention to stress is prudent. In the chemical sector, safety is paramount; creating a safety climate and culture is critical to reducing accidents in the workplace. Human factors play a pivotal role in the success or failure of a safety culture. An astonishing 80% of workplace incidents are a result of stress related factors. In this session, we'll explore how measuring and enhancing employee personal resilience can reduce the likelihood of stress related errors and buy into the wider safety program. We'll also explore the benefits of a personal resilience program and discuss stress coping strategies and resilience boosting skills, along with holistic suggestions for tackling the invisible hazard in your organization.

168b When Safety Becomes Unsafe: A New Model for Proactively Improving Human Performance

David Wilbur

With incidents on the rise and traditional root cause analyses no longer producing desired results, a new model for analyzing the human performer in incident investigation is needed to break through the cycle of repeated costly mistakes and performance plateaus.

In this interactive session, we discuss the need to shift from traditional causal models of human performance to an evolved adaptive performance approach, accomplished through building operational resilience.

Operational resilience is defined as the capacity of an organization's people to adapt their responses to changing system conditions through normal decision making to produce desired safety and business outcomes. Cultivating adaptive capacity is the critical enabler of building operational resilience and the human performer's ability to proactively identify and respond to normal and unanticipated system dynamics, emerging risks, and navigate disruptions.

Using both qualitative and quantitative methods, we developed and evaluated a multi-dimensional model of operational resilience measuring the interactions between operating discipline (mechanical, process and human reliability), latent environmental factors, and multiple dimensions of adaptive capacity. Examination of the complex relationships of human performance within a dynamic systems environment, gives us insight into blind spots and opportunities to develop the ability of leaders and operators to anticipate change, learn from experience to deal with changing conditions, monitor systems for imbalance, and adjust organizational processes to effectively navigate and mitigate risks.

By illustrating the relationships between operating discipline and adaptive capacity, participants will have insight into the critical human performance strategies needed to develop practical solutions valuable in their safety and business operations at home.

181e Investigation Findings and Lessons Learned from the 2012 MSC Flaminia Explosion and FIRE

Scott Davis, Derek M. Engel, Tom DeBold, John Pagliaro

On July 14, 2012, an explosion and ensuring fire significantly damaged the container ship MSC Flaminia hat was transporting cargo from the United States to Europe. The vessel was first loaded in New Orleans 14 days prior to the incident and had made subsequent port calls to finish cargo loading, where it began its transatlantic voyage on July 8, 2012, with over 4,800 containers stowed on and below deck. Early on the morning of July 14, smoke alarms began to sound in cargo hold no. 4 and a white cloud was visible emanating from the hold. The crew incorrectly assumed that there was a fire in hold no. 4, hence began CO_2extinguishing activities and ultimately entered hold no. for further firefighting activities. Their activities likely initiated an explosion that killed three crew members and resulted in a subsequently fire that damaged thousands of containers.

All parties agreed that the incident was caused by a runaway condition from the auto-polymerization of 80% grade of divinylbenzene (DVB80) that was contained in ISO-containers within hold no. 4. DVB80 is a monomer that can exothermically form a polymer and the energy generated by this reaction can rapidly increase the polymerization rate (self-promoting and auto-accelerating), until it can no longer be controlled, or runaway occurs. Inhibitors in conjunction with dissolved oxygen are added to DVB80 to prevent this auto-polymerization. This paper presents the results of an extensive investigation into whether the runaway was cause by insignificant oxygenation during DVB80 manufacture or was it caused by the storage conditions at the port and on the ship.

181c The Day an HF Release Caused a Serious Operator Injury - Lessons Learned from Incident Investigation

Peter Mcknight, Ronald Melucci, Dalia Davila

Results of an Incident Investigation following an onsite anhydrous hydrofluoric acid (HF) release with extreme impact to an operator ranks as one of the major events in my career. It confirmed how vital application of PSM is to prevention of and preparedness for response to high hazard loss of containment scenarios. The thorough work completed by plant personnel leading up to the day the HF release and operator injury occurred included a thorough Management of Change (MOC) process for installing a new sight glass in an HF line. The Author was personally engaged in the MOC process, the Pre-Startup Safety Review and all system and field checks prior to charging the line with HF. He was present when the release occurred and observed the emergency response as well as the medical procedures used to save the operator's life (with multiple calcium gluconate injections, decontamination and stabilization of the patient) prior to transfer to the hospital.

Even when PSM is fully implemented properly, there are still factors that can lead to loss of containment and operator injury. This event happened early in my career, cemented the need for thorough PHA, MOC and PSSR as well as Incident Investigation to learn from the incident. Dealing with an extremely hazardous substance such as HF during a loss of containment event and having trained responders, procedures, supplies and trained medical staff ready to respond to HF contamination and its toxic effects. This incident and the HF Plant Team's response to it; that resulted in saving the operator's life, left an indelible lesson in my mind. The incident investigation evaluated all contributory factors and determined the root cause of failure. This drove home why PSM matters so much to me while implementing PSM across the industry.

181a Enhancing ISO Container Loading Design for Chemical Transport

Ali Bazroun, Hashem Al-Awami

The safe and efficient loading of ISO containers for chemical transport is a critical yet challenging aspect of the chemical industry. However, there has been a notable absence of clear guidelines and standards in this domain, necessitating the development of comprehensive design practices. This paper addresses this gap by presenting a novel approach to ISO container loading design, with a specific focus on minimizing human error, high-pressure situations, overfill incidents, and spill containment.

This study was initiated in response to a chemical loading project undertaken by Sabic, which underscored the urgency of establishing robust design practices. The practices were developed through an extensive process that involved benchmarking existing facilities and consulting Sabic process designers. Furthermore, a thorough Process Hazard Analysis (PHA) was conducted to evaluate the risk level associated with the loading activity and assess the effectiveness of safeguards.

One of the distinctive features of this paper is its in-depth examination of the role of human intervention in the loading process. The research emphasizes the significance of aiding operators through optimal design, reducing the potential for human error, and enhancing overall safety. Through a combination of case studies and analysis, this paper provides insights into the root causes of human error during ISO container loading and offers practical recommendations for minimizing its occurrence.

By shedding light on the crucial aspects of ISO container loading design and human factors, this paper contributes valuable knowledge to an industry grappling with safety concerns and regulatory compliance. The proposed design practices aim to enhance operational safety, reduce risks, and improve the overall efficiency of ISO container loading in chemical transport, ultimately benefitting both workers and the environment.

SPRING24
+20TH GCPS
A Joint AIChE and CCPS Meeting

Implementing A Simultaneous Operations (SIMOPS) Process During the Superior Refinery Restart

Ola Olamilehin
Cenovus Energy
225 6 Ave SW, Calgary, AB, T2P0M5
Ola.Olamilehin@cenovus.com

Brian Steinke
Cenovus Energy
225 6 Ave SW, Calgary, AB, T2P0M5
Brian.Steinke@cenovus.com

Prepared for Presentation at
American Institute of Chemical Engineers
2024 Spring Meeting and 20th Global Congress on Process Safety
New Orleans, LA
March 24-28, 2024

Implementing Simultaneous Operations at Superior Refinery

Ola Olamilehin
Cenovus Energy
225 6 Ave SW, Calgary, AB, T2P0M5
Ola.Olamilehin@Cenovus.com

Brian Steinke
Cenovus Energy
225 6 Ave SW, Calgary, AB, T2P0M5
Brian.Steinke@cenovus.com

Keywords: SIMOPS, Simultaneous, Operations, Superior, Refinery, Process, Implementing

Abstract

After a fire incident at the Superior Refinery Wisconsin in 2018, a rebuild project was executed to restore the facility. The refinery has a processing capacity of 49,000 barrels per day (bbls/d) and can produce a full slate of products, including asphalt, diesel, gasoline and heavy fuel oils. A restart plan that addressed processing unit startups during simultaneous construction and commissioning work led to the implementation of a Simultaneous Operations (SIMOPS) Process which used an activity matrix tool for decision making. The SIMOPS activity matrix was developed with consideration of operational, construction, and commissioning activities. Each activity scenario on the matrix was than assessed and categorized as acceptable, not acceptable, or restricted. A SIMOPS Person-in-Charge (PIC) was identified to implement the SIMOPS process during the construction, commissioning, and process unit startups. The implementation of the SIMOPS process allowed work activities to safely continue while starting up process units, thereby optimizing the refinery restart, and leading to early production of asphalt and diesel products.

1 Implementing a Simultaneous Operations Process during the Superior Restart

The Superior Refinery Wisconsin underwent a rebuild project after experiencing a fire incident in 2018. The rebuild project restored processing capacity for a full slate of hydrocarbon products.

The refinery restart involved startup of several process units either in sequence or together. Typically, all construction and commissioning work would be complete, and operations would focus on safely starting up the facility while minimizing non-essential personnel.

To optimize the restart process for the Superior refinery, the rebuild project construction and commissioning work activities continued during process unit startups. This resulted in significant numbers of construction and commissioning workers on-site during the process unit startups requiring additional mitigations to manage the simultaneous operations.

A restart plan that addressed processing unit startups during simultaneous construction and commissioning work led to the implementation of a Simultaneous Operations (SIMOPS) Process.

2 Optimization Considerations

Just like any other optimization process, there are potential challenges that need to be considered. In this case, the challenges and complexities are revealed by looking at the project statistics.

2.1 Project Challenges

Overall project statistics below show the activities, personnel, and scale of the rebuild project, and the need for a process to manage the complexity of the startup.

- Project duration was from 2019 through 2023
- Over 12 million person hours worked
- 2200 peak construction workers (August 2022)

2.2 Challenges to Solution

Optimizing the refinery restart plan meant minimizing interruption for on-going construction work within several process units in the facility. The first process unit startup was the introduction of diesel (hydrocarbon) into the crude unit. The challenge was maintaining a safe working environment during the introduction of the hydrocarbon hazard. A solution was developed to maintain a safe working environment, progress the construction, and carry out the refinery restart plan.

Based on leadership experience using SIMOPS processes to safely manage complex off-shore operations, a proposal was made to adopt a similar approach for the refinery restart.[1] In the past, SIMOPS was not fully integrated into the safe control of work practice during a refinery startup. Therefore, a SIMOPS Package was developed that safely managed the simultaneous work process.

3 Overview of the SIMOPS Package

As part of managing activities for the restart, refinery leaders were introduced to the concept of SIMOPS which includes the following components:

- SIMOPS Activity Matrix
- SIMOPS Person in Charge
- SIMOPS Chain of Command
- SIMOPS Plan

4 Implementing the SIMOPS Process

Each component of the SIMOPS package works together to formulate a complete SIMOPS process. Development, assignment, training, and planning activities are performed to support the implementation work required to manage the SIMOPS process.

4.1 SIMOPS Activity Matrix Development

The SIMOPS Activity Matrix is a decision making tool to assess the safe control of daily work activities occurring simultaneously.

The SIMOPS Activity Matrix for the refinery was developed over multiple workshops to ensure the matrix was usable for its purpose. Preparation for the workshops included two important criteria:

i. Identify the subject matter experts at the refinery to provide the inputs for the SIMOPS Activity Matrix.
ii. Develop the sequence of commissioning and startup activities.

Development of the SIMOPS Activity Matrix identified over 150 activities that created over 10,000 potential SIMOP scenarios that required evaluation.

The SIMOPS Activity Matrix was created in three basic steps[2]:

i. SIMOPS Activity Identifications
ii. SIMOPS Activity Assessments
iii. SIMOPS Activity Restrictions

Figure 1 shows an example of the SIMOPS Activity Matrix developed for restarting the refinery.

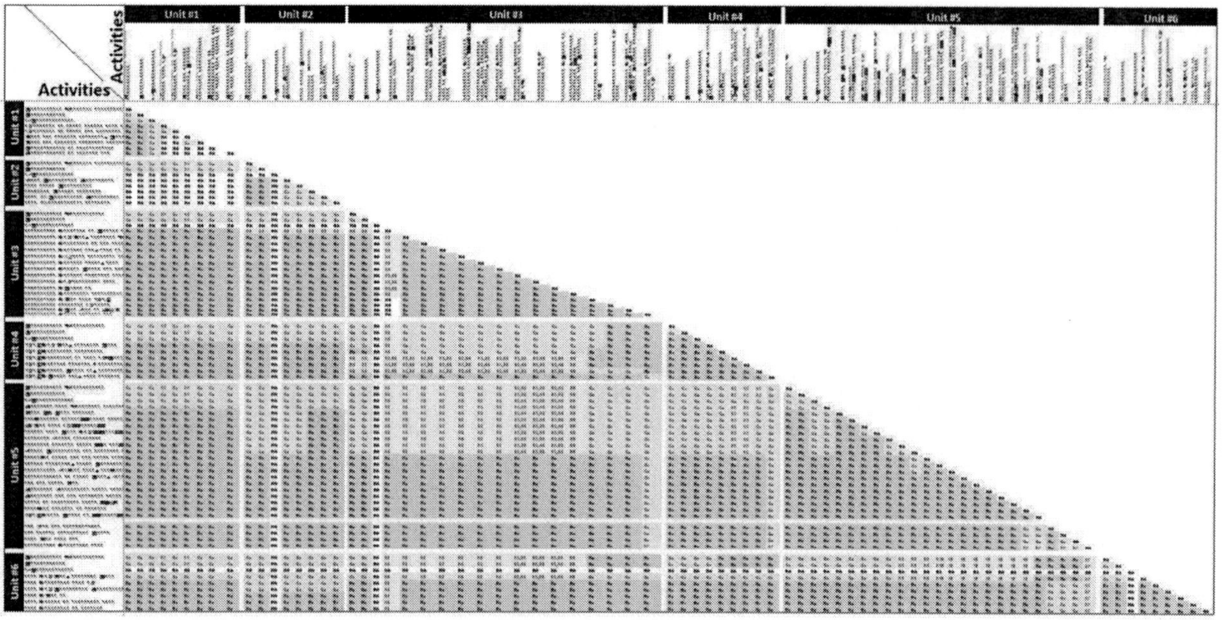

Figure 1: Example of the refinery SIMOPS Activity Matrix (Un-readable due to proprietary content).

Activities were listed on the first column and the top row in the same order. In this example, activities were additionally grouped by process unit number in which they occurred in the startup sequence. Activity assessments were then categorized and represented in the colored sections.

4.1.1 SIMOPS Activity Identifications

The first step in the matrix development was to identify the work activities that may occur simultaneously during the construction, commissioning, and operational startup steps.[3]

The activity identification step required consideration of the following subjects:

i. Identifying the type of activity (examples as follows)
   - N2 purging
   - Insulation of piping
   - Circulating gas oil
ii. Identifying the functional groups (examples as follows)
   - Operations
   - Construction
   - Commissioning
iii. Identifying the hazards includes
   - Hazards associated with at least one activity such as nitrogen purging of a line. For example, if there is a leak of nitrogen during purging of a line while conducting insulation work or other types of work could lead to asphyxiation of workers in the area.

- Hazards associated with two or more activity occurring at the same time such as circulating gas oil while welding in the area. In this scenario, if there is a leak from circulating the gas oil while performing welding, a fire risk can occur.

4.1.2 SIMOPS Activity Assessments

The second step in the matrix development was activity assessment, where simultaneous activities were assigned to categories when mapped against each other. Four categories were used and each was given a traffic light color coding to assist in the readability of the tool. The four categories are listed in Table 1.

Table 1 Simultaneous Activity Categories

Work Category	Code	Meaning
Standard Functional	Go (green)	Normal permitting and other controls are in place for two or more concurrent activities
Restricted	R# (amber)	Allows two or more concurrent work activities with specific controls put in place to safely manage the simultaneous work
Not Permitted	No (red)	No simultaneous activity allowed
Not Applicable	N/A	Two or more activities do not conflict in any way

Categorization of the activity resulted in some temporarily un-categorized activities. These required further study utilizing either a high- or detailed-level risk assessment before the workshop team could categorize them accurately. In some cases, consequence modelling was required to categorize the activity.

4.1.3 SIMOPS Activity Restrictions

The third step in the matrix development focused on what restrictions or control of work measures were required that would safely allow these activities to occur simultaneously, if possible. A hierarchy of controls approach was used to establish the restrictions that would allow the work to proceed.[4]

Hierarchy of controls methodology was used for assessing the effectiveness of a restriction when evaluating the risk controls required to keep the simultaneous activities safe.[5] A representation of this hierarchy of controls approach is found in Figure 2.

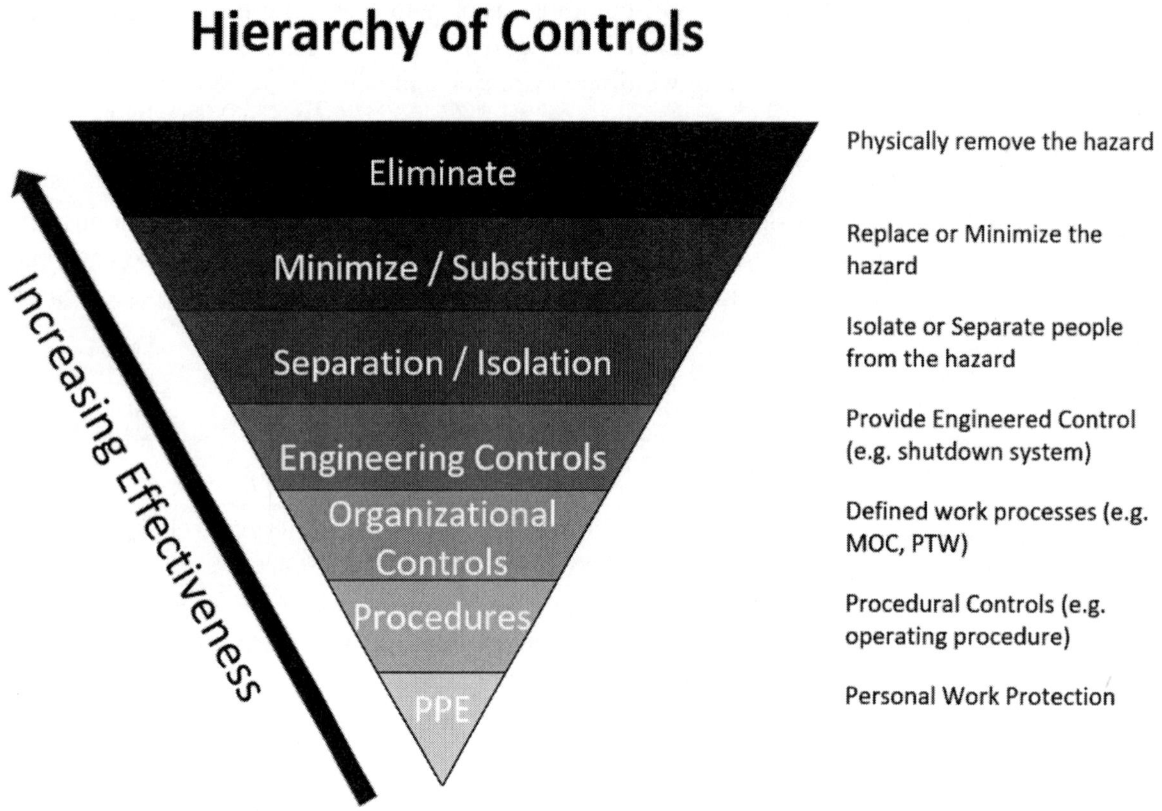

Figure 2: Illustration of a hierarchy of controls methodology for assessing the effectiveness of a restriction.

Some examples of the types of restrictions used to control the work came in the form of:

- Equipment removals
- Minimum manning
- Exclusion zones
- Barricades
- Spotters/Safety Watch

4.2 SIMOP PIC

After completion of the refinery SIMOPS activity matrix, the concept of the SIMOPS Person-in-Charge (PIC) was established to implement the rest of the SIMOPS components.[6]

In the simplest terms the SIMOPS PIC was the "Gatekeeper" of the SIMOPS Activity Matrix. The SIMOPS PIC managed the matrix, oversaw its implementation, and utilized the tool every day.

The SIMOP PIC was a dedicated resource that focused only on simultaneous operations and the hazards associated with those activities. All other roles within the project and restart plan were well-established.

The SIMOPS PIC responsibilities included:

- Ensure compliance to all categories and restrictions listed on the SIMOPS Activity Matrix and with the Permit to Work system.
- Act as a single point of contact for all simultaneous operations including change in sequence or change in SIMOPS criteria.
- Lead SIMOPS discussion in the War room (communication room for all stakeholders), ideally every morning and at the start of any SIMOPS activity
- Log, review, and accept all SIMOPS activities, before commencing the SIMOPS activity.
- Complete field verification that SIMOPS activity matrix restrictions are in place prior to commencing the SIMOPS activity.

The SIMOPS PIC authority was established by Management after approval to implement the SIMOPS process and the SIMOPS PIC had complete authority to determine which operation or phase of the work had precedence at any given time.

4.3 Establishing the SIMOPS Chain of Command

The SIMOPS PIC used the SIMOPS Activity Matrix tool to communicate impacts of simultaneous work activities by consulting with appropriate authorities to make decisions on work that is to be completed every day. Those involved in the daily work activities from leader or coordinator level would participate in these meetings. Figure 3 shows the organization structure for the refinery SIMOPS chain of command process.

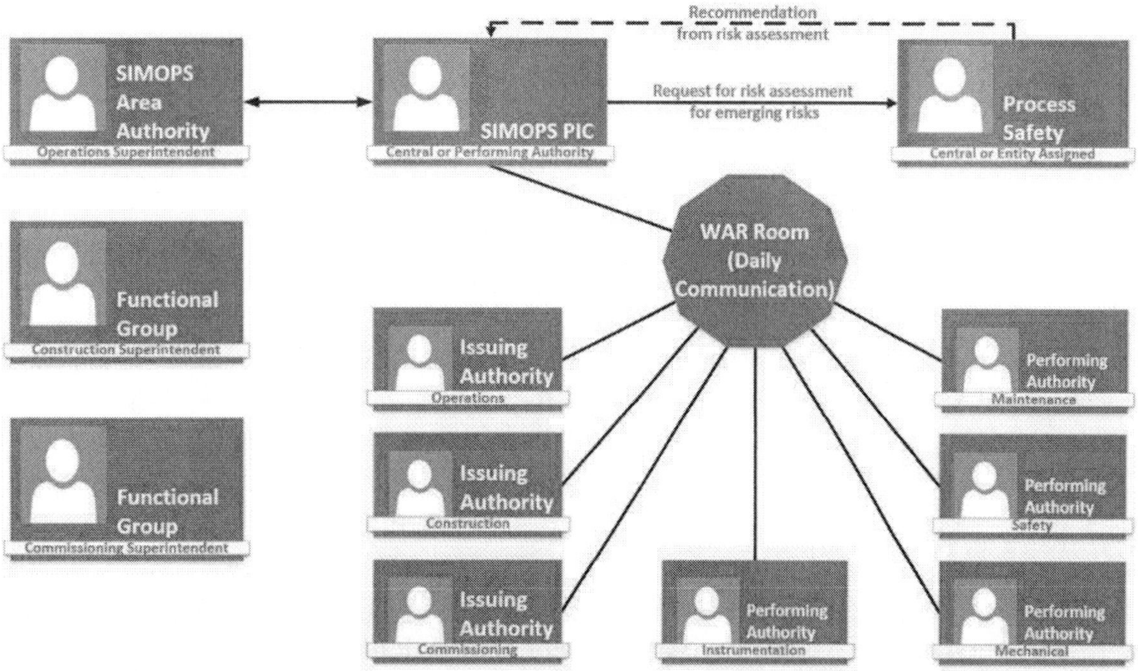

Figure 3: Refinery Organization Structure for SIMOPS Chain of Command.

The daily communication meetings, often referred to as the "War" room meetings were intended to be short update meetings on the daily activities that would be checked against the SIMOPS Activity Matrix for quick identification of any concerns.[7]

A single area authority who was responsible for the highest risk tasks had the final authority on SIMOPS activities, but the SIMOPS PIC using the SIMOPS Activity Matrix tool made most of the decisions during activity evaluation. If concerns were raised as to the categorization of the SIMOPS activity, resolution meetings were held to re-assess the categorization.

Ensuring an adequate level of assessment in the development of the SIMOPS Activity Matrix will minimize these types concerns along with backup documentation of why the categorizations were selected.

4.3 SIMOPS Plan

Development of the SIMOPS plan required that SIMOP initiators like commissioning and startup or operations personnel must communicate SIMOPS plans to all stakeholders, and review plans with the SIMOPS PIC, facility supervisors, and construction representatives involved in the work.

An important aspect to the SIMOPS plan was that the SIMOPS PIC accepted the proposed activities before commencing the SIMOPS activity. The SIMOPS plan also encouraged maintaining the same Operators, Maintenance, and Project/Construction personnel through the entire SIMOPS activity.

SIMOPS planning included:

- Written scope of work.
- Completion of Safe Job Analysis, Field Level Hazard Assessments or Equivalent.
- Up to date (or marked up) job specific process safety information necessary to accurately identify workplace hazards during SIMOPS planning, approvals and execution (e.g. PFD, P&ID, Plot plan, etc.).
- An emergency response plan, including clear lines of communication and emergency response responsibilities.
- Approved permit to work issued through operations permit to work system, or equivalent.

4.4 Keys to SIMOPS Implementation Onsite

Implementing a SIMOPS process to stakeholders required facilitating some training and education on the what, why, and how of the SIMOPS process. After the stakeholders understood the safety benefits of the SIMOPS process, acceptance of the implementation was achieved. To manage the implementation, the SIMOPS PIC focused on the following:

- Establishing relationships between various stakeholders to ramp up the SIMOPS activities.
- Recognizing the importance of providing clear and concise direction to the workforce.
- Making time to communicate with field leaders.
- Coordinating with stakeholders to setup SIMOP restrictions (exclusion zones, barricades, spotters, etc.).
- Resolving SIMOPS conflicts or concerns with the identified appropriate authorities.

5 Benefits of the SIMOPS Process

The refinery operations restart plan benefited in many ways by supporting and implementing the SIMOPS process into their activities. Some of the following benefits links directly to the SIMOP process:

- Progressing construction work activities safely while starting up process units. This was achieved because of the work done to develop the SIMOPS Activity Matrix and implementing restrictions to safely perform the simultaneous activities.
- Optimization of the refinery restart was possible because the SIMOPS process allowed for construction work activities to safely continue while starting up each process unit.
- Communication to functional groups through the daily War room meeting.
- Assignment of a SIMOPS PIC to implement restrictions through the SIMOPS chain of command and control of work process.

6 Learnings during the SIMOPS Implementation Process

Socialization of the SIMOPS process with the refinery stakeholders helped to gain alignment with construction and commissioning groups. However, earlier engagement will further increase awareness.

High level risk assessment was used to categorize the activity while developing the SIMOP Activity Matrix. Formal risk assessment of each potential SIMOP scenario will reduce the possibility of deviations from the matrix.

Identifying restrictions was done during the development of the SIMOPS Activity Matrix. Implementation of the restrictions helped to control the hazards that existed for the onsite construction and commissioning workers. However, developing a restriction implementation plan would further improve the general risk awareness of the simultaneous activities that could lead to an incident.

Completing each component of the SIMOPS package successfully contributed to the implementation of the SIMOPS process. Additionally, embedding the SIMOPS implementation package in the restart plan will contribute to an effective execution of the SIMOPS process.

7 Results of SIMOPS Process Implementation

The results of implementing a SIMOPS process to allow for a safe optimized restart of the refinery was considered successful in that:

- No incidents were recorded resulting from simultaneous activities.
- Optimizing the restart of the refinery was achieved, leading to early production of asphalt and diesel products.

Factors that contributed to achieving the above results are:

- The SIMOPS Activity Matrix that provided daily decision-making capability.
- A Chain of Command process that included a dedicated SIMOPS PIC, provided an acceptable approach for the SIMOP process implementation.
- Implemented restrictions that were identified on the SIMOPS Activity Matrix.
- Daily communication in War room meetings encouraged resolution to simultaneous activity concerns.

The SIMOPS process will improve the safety around complex and challenging work activities. The process has flexibility to be scaled to suit other projects and facility turnarounds. Implementing this concept will contribute to a safe working environment and should be seen as a standard component of a well planned start-up / turnaround.

8 References

[1] Singh Barinder. Simultaneous operations in oil and gas installations and work sites; International Journal of Innovative Science, Engineering & Technology 2016; 3:19-23

[2] Chapter 9 Maintenance, Sutton Ian Plant Design and Operations, Gulf Professional Publishing, 2017, pp. 265-292

[3] Energy Institute Guidance on Managing Process Safety in Decommissioning Projects 3.6.2.7 SIMOPs – Knovel. Accessed 24 January 2024

[4] BP US PIPELINE & LOGISTICS Level 2 HITRA procedure 2020 https://www.bp.com/content/dam/bp/country-sites/en_us/united-states/home/documents/products-and-services/pipelines/contractor-information/policy/level-2-hitra-policy.pdf Accessed 24 January 2024

[5] Canadian Centre for Occupational Health and Safety. Hazard and Risk. https://www.ccohs.ca/oshanswers/hsprograms/hazard/hierarchy_controls.html . Accessed 22 February 2024

[6] BP WIND ENERGY Simultaneous Operations 2011 https://www.usbr.gov/lc/region/g2000/envdocs/MohaveCountyWindFarm/Plan_of_Development/508%20Attachments%20to%20HSSE%20Plan/Attach_NN_SIMOPS.pdf Accessed 24 January 2024

[7] Chapter 30 Simultaneous construction and commissioning considerations, Killcross, Martin. 2021 Chemical and Process Plant Commissioning Handbook - A Practical Guide to Plant System and Equipment Installation and Commissioning (2nd Edition). Elservier, pp. 231-233

111g Risk Assessment Model for Defining Minimum Distance of Shipping Lanes from Offshore Wind Farm Location

Mariana Bahadian Bardy, Bruno Wehmuth, Cynthia Spitzenberger

With the rush of Energy Transition, Offshore Wind Farms are rapidly becoming a greater contributor to the energy grid, as presented by DNV in the 2023 Energy Transition Outlook (https://www.dnv.com/energy-transition-outlook/index.html). For its development, several wind farms are planned in waters with high shipping density. This may create a potential conflict of interest between navigation of ships and the wind farms due to different operational aspects. When a wind farm is installed, safety distances from wind farms will be defined and this implies in modification of shipping lanes, that will interfere on commercial shipping. It can also influence on fishing ships, for the farms located in shallow waters. Additionally to that, an increase on shipping traffic is expected based on requirements for installation, maintenance and supply for wind farm operations.

Although a collision is an undesired event for both segments, the understanding of impacts to business continuity and in terms of who is responsible for tanking the proper measures to reduce the risks sometimes is not entirely clear. These uncertainties can escalate turning in a scenario of implementation of restrictions by regulatory agencies. This scenario can delay the installation of new wind farms and therefore delaying the energy transition.

The best way to understand the possible impacts and also to promote important discussions about the subject is to gather the maximum information about the risks associated with using a formal safety assessment to ensure that all actions will be taken to reduce the probability of collision.

The safety assessment will help to understand the expected consequences from a ship collision against a wind turbine offshore, as:

- production loss due to impact on offshore transformer;
- total loss of ships and loss of life due to collision of smaller ships (fishing ships, pleasure boats);
- ship impact against manned installation, with structural loss and potential fatalities;
- potential serious damage to ships and multiple fatalities in case of impact of passage ships;
- potential oil pollution.

The objective of this paper is to present a model to estimate shipping collision frequency and shipping collision impact towards offshore wind farms, with focus on merchant passing vessels, considering different volumes of shipping traffic and shipping size versus distances from shipping lanes.

This paper explore methodology for ship frequency calculation, based on DNV Recommended Practice DNV-RP-F107 and the methodology presented in The Centre for Marine and Petroleum Technology (CMPT), A Guide to Quantitative Risk Assessment for Offshore Installations, 1999, that define how to calculate the likelihood of having a ship collision of a passing vessels against an offshore platform. This methodology has been adapted by authors to estimate frequency of collision against wind farm turbines installed offshore.

This paper will calculate two events of collision, powered collision where Merchant vessels can collide against wind farms with powered engines or they can drift-off and be trawled by the currents, waves and wind towards wind farms.

For this paper, model is adapted to use data extracted from AIS, regarding density (number of ships passing), ship velocity, and typical ship size.

Results are presented as expected number of incidents per year, and expected design accidental load to support decision on possible mitigations as:

- Location of Offshore Wind Farms;
- Definition of collision load for design of offshore structures;
- Additional mitigation for reducing likelihood of ship collision.

Additionally, different distances and volumes of traffic are assessed in order to defined tolerable distance from shipping lanes, for several volumes of traffic. This information is relevant to support decision of location of offshore wind farms, focus in minimizing impacts of shipping collision, especially for areas located near high traffic areas.

193a How AI and Machine Learning Can Improve Process Safety Performance

Jimmy Miller

Since 2020, the trending focus in the process automation and safety industry has been Digital Transformation. While this concept has gained traction and a large portion of our industry truly sees the value that can be realized by beginning their digital transformation journey, many of these efforts have fizzled due to the difficulty in harnessing large volumes of data into actionable insights. The utilization of artificial intelligence (AI) and machine learning (ML) has gained significant attention in recent months. The use of AI and ML algorithms can provide valuable insights into safety hazards and risk mitigation strategies, leading to improved safety performance and reduced accidents in industrial processes.

This paper explores the value of utilizing AI and ML in improving industrial process and functional safety. It discusses how these technologies can be used to identify safety risks, predict potential hazards, and optimize safety measures. Additionally, it explores the benefits of AI and ML in automating safety management tasks and reducing human error.

Overall, this paper highlights the potential of AI and ML in improving industrial process and functional safety and provides an overview of their benefits and limitations. It concludes that the integration of AI and ML into process safety management systems can lead to safer working environments, increased efficiency, improved maintenance procedures and reduced costs.

Good Practices for Effectively Managing Process Safety Issues When Replicating a Modular Mid-Scale LNG Design

Author: Luna, Veronica Christine B., HSE Director (Process Safety), Fluor Corporation, Philippines

Presenter: Riffer, Ryan, HSE Director, Fluor Corporation, Houston, USA

Abstract:

Modular mid-scale LNG is emerging as one of the potential solutions to meet the needs of the evolving LNG market and global demand for clean energy. It offers a faster and cheaper LNG liquefactions plants than conventional plants. The modular solution offers opportunity to decrease time to market for the clients, as well as lower total installed cost. It also has the flexibility of being installed in different locations and different configurations, in a plug-and-play manner.

When implementing this solution on a project, a reference design is available as a starting point. This can then be modified to meet specific requirements for a certain client. Once the reference design is 'customized' and made client-specific, it can then be replicated with the purpose to meet the total plant capacity which in conventional practice, being achieved by a single large-scale plant

The customized base design can also be replicated with the intention to either be installed in a later phase in same facility for capacity expansion or for installation in other locations or another type of facility. For these cases, the expectation is that there will be minimal efforts in design and procurement during the replication process.

As it is in any other copy job, there is a misconception that all process safety features of the base design can be carried over when arranged differently or installed in another location or in a different type of facility. However, although the inherent process safety hazards associated with the base design may not change for most of the process units, not all process safety features can be copied as other considerations must be accounted for when there is a change in location, environmental conditions, regulatory requirements, or facility general arrangement.

This paper shares practices to effectively manage process safety issues when replicating a modular mid-scale LNG design and is based on recent project experiences. It presents the inherent process safety hazards associated with the base design, discusses the process safety features which are not expected to change and highlights other considerations when the location and/or facility general arrangement is changed.

The paper will also present some good practices on how process safety efforts can be minimized during replication process without compromising quality of the outcome of process safety activities and while maintaining the focus of working towards the goal of providing a compliant and safe design to our clients who will ultimately operate the facility.

Keywords:

Modular_mid-scale_LNG, process_safety_in_design, liquefaction

1.0 Introduction

Modular mid-scale liquefied natural gas (LNG) is emerging as one of the potential solutions to meet the needs of the evolving LNG market and global demand for clean energy. It offers an opportunity to decrease time to market for the clients, as well as lower total installed cost (TIC) which translates to a faster and cheaper LNG liquefactions plants than conventional plants.

A reference design is available as a starting point which can then be customized to meet specific requirements for a certain operator which will be referred to as the client in this paper. Similarly, this customized reference design will be referred to as the "base design" throughout this paper.

Once a reference design is customized and made client-specific, it can then be replicated with the purpose to meet the total plant capacity which in conventional practice can be achieved by a single large-scale plant. This solution provides an option for clients to stage their investment while generating revenue. It also has the flexibility of being installed in different configurations in a plug-and-play manner. The customized base design can also be replicated with the intention to install the modules in other locations or another type of facility (e.g. land or marine based). For these cases, the expectation is that there will be minimal efforts in design and procurement during the replication process.

This paper shares good practices for effectively managing process safety issues when replicating a ready-made design like the modular mid-scale LNG solution. It will also present some recommendations on how process safety efforts can be minimized during replication process without compromising quality of process safety activities and maintain the focus of providing a compliant and safe design to the clients who will operate the facility.

1.1 Overview of Modular Mid-Scale LNG Solutions

Large-scale liquefaction plants have capacities between 3.5 and 8.0 Million Tons Per Annum (MTPA) while mid-scale LNG facilities produce trains in 0.7-3 MTPA range (Patel et al., 2021). This mid–scale LNG design utilizes one of the liquefaction technologies available – single mixed refrigerant (SMR), propane pre-cooled mixed refrigerant (C3MR) or dual mixed refrigerant (DMR) depending on the required train capacity.

Reference designs were developed with the aim to provide starting points for future EPC projects where mid-scale LNG solution is to be utilized. This expedites project delivery as it requires a shorter time for conceptual phase and front-end engineering design (FEED). Typical deliverables produced as part of the reference design development effort are process flow diagram (PFD), heat and material balances (HMB), equipment datasheets, sized equipment list, utilities summary, piping and instrumentation diagrams (P&IDs), preliminary plot plan, preliminary three-dimensional (3D) model, and modular execution philosophy. The layouts were based on maximizing modular content and minimizing the work to be executed at site while still considering industry standards in terms of equipment spacing for safety, material handling, operations and maintenance. Rendition of the 3D model of reference design examples are shown in **Figure 1** (Patel et al., 2021).

Figure 1. LNG Mid–Scale Facility with Individual SMR LNG Train – Onshore versus Offshore (Patel et al., 2021)

The overall facility reference designs considered minimal initial overall facility needs and including the following units:

- Inlet gas receiving
- Feed gas pre-treating (includes mercury removal, acid gas removal, and gas dehydration)
- Heavy hydrocarbon removal
- Liquefaction train
- Refrigerant storage
- Boil-off gas handling
- Utilities and offsites (fuel gas, air, power generation, flare, etc)
- LNG storage
- Ship loading and marine facilities

The reference designs were based on a pipeline feedstock that is readily available and climatic conditions that are representative of multiple locations. Although developed with land based installations in mind, the designs could be adapted to offshore marine environments where size and weight are major considerations.

1.2 Process Safety Features Incorporated in Reference Design

Process safety hazards inherent to the process technologies applied in each reference design were evaluated during the development stage. The hazards identification step was initiated by the review of the HMB, PFDs, equipment list and safety datasheets (SDS) of the substances involved. The hazards associated with the mid-scale LNG reference design considering an SMR liquefaction technology are listed in **Table 1**. In general, these hazards will not change as long as feed gas composition and process technology remain the same. However, the magnitude of consequences and associated risk reduction measures can differ for each facility.

Table 1. Hazards associated with Recently Developed Mid-Scale LNG Reference Design

Material/Substance Involved	Associated Hazards	Associated Consequences
Natural Gas (NG), mostly methane	Flammable gas	Fire, vapor cloud explosion (VCE)
LNG, mostly methane (at cryogenic temperature)	Flammable liquid/vapor Operating at cryogenic temperature Asphyxiant	Fire, VCE, boiling liquid expanding vapor explosion (BLEVE) Personnel/asset exposure to cryogenic temperature Personnel exposure to oxygen deficient atmosphere
Refrigerant (propane, butane, pentane, etc. at low temperature)	Flammable liquid/vapor Operating at low temperature Asphyxiant	Fire, VCE, BLEVE Personnel/asset exposure to low temperature Personnel exposure to oxygen deficient atmosphere
Heavy Hydrocarbons	Flammable liquid (C5 and heavier)	Fire, VCE
Hydrogen Sulphide	Acute toxin	Personnel exposure to acute toxic gas
Heating Medium (Hot Oil)	Combustible/flammable liquid	Fire
Amine	Corrosive to skin Toxic to environment	Personnel exposure to harmful substance Adverse impact to the environment
Carbon Dioxide	Asphyxiant	Personnel exposure to oxygen deficient atmosphere

Safety features integral to the process design such as application of inherently safer design (ISD) concepts, basic process controls including alarms and typical safety instrumented functions can be incorporated during the preparation of the P&IDs and specification of equipment. However, process safety features and risk reduction measures that can be incorporated in the reference design are limited since they heavily depend on factors which will be further discussed in this paper. Management of process safety risk will need to be specific for each facility and associated location.

2.0 Risk Management During Base Design Development and the Replication Process

2.1 Overview of the Risk Management Process

As a precursor to the discussions about how process safety issues can be managed when replicating a modular mid-scale LNG design, it is important to review the framework to be applied in the process. Doing so provides a guidance map to maintain the focus towards achieving risk reduction to tolerable levels. As seen in **Figure 2**, the risk management process (RMP) begins with understanding what drives it. This requires having a good grasp of the requirements imposed by local regulators, those stipulated on applicable codes and standards, the client-specific requirements including their risk tolerability criteria and can include other stakeholders' such as insurer. This defines the minimum expectations and the facility's risk reduction targets. Risk assessment is the heart of the risk management process which is the key step in achieving tolerable risk levels. Facility location, process design, and plot arrangement are the critical input data for performing this step. Any changes in the minimum requirements and information needed for the risk assessment alters how risk is to be treated and the risk reduction features to be implemented.

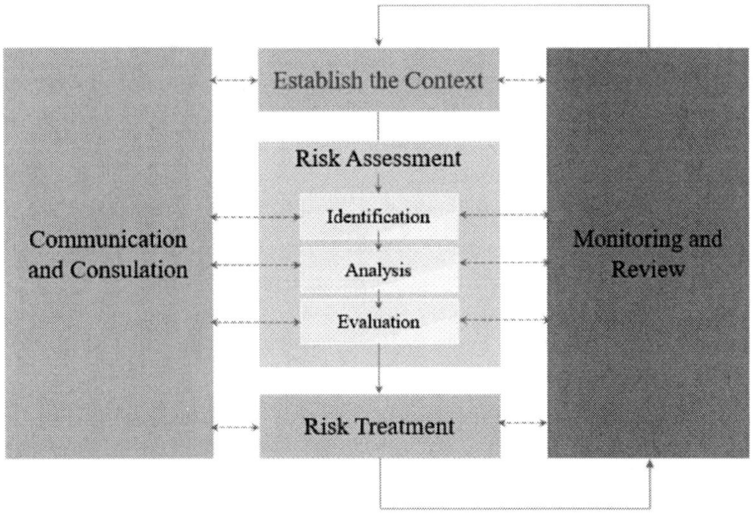

Figure 2. Risk Management Process (IOGP, 2018)

2.2 Incorporating the Risk Management Process in Base Design Development

For most clients, activities leading to finalization of the base design and its construction are performed utilizing a phased approach in project execution. Design stage is subdivided into conceptual phase, front-end engineering design and detailed design. **Table 2** summarizes the typical process safety related outputs in each phase.

Table 2. Typical Process Safety Related Activities during Base Design Development per Design Stage

RMP STEP		CONCEPT SELECTION	FRONT-END ENGINEERING DESIGN	DETAILED DESIGN
Communication and Consultation	**Establish the Context**	Local regulations review Project requirements established	Local regulations review Project requirements finalized	Local regulations review Project requirements incorporated
	Risk Assessment	HAZID/ENVID workshops Conceptual HSE studies, as required	Preliminary HSE studies Preliminary PHA and SIL Assignment	Updated HSE studies, as required Detailed HAZOP and SIL assignment including vendor packages
	Risk Treatment	Preliminary HSE philosophies Engineered risk reduction measures identified Preliminary plot layout Customized reference design	Finalized HSE philosophies Specification and basic design of engineered risk reduction measures Updated plot layout Basic design finalized	Detailed design and procurement of engineered risk reduction measures Plot layout for construction Detailed design finalized
	Monitoring and Review	Inter-disciplinary checks	Design reviews/workshops Management of change, if required HSE verification activities	Design reviews/workshops Management of change HSE verification activities

Considering the information available in the reference design, it is expected that shorter duration and reduced efforts are required to complete the design of the modular mid-scale LNG facility. However, this is not applicable for process safety activities since minimal risk management efforts can be accomplished as part of the reference design development.

To support the accelerated project schedule, strategic planning of how activities will be executed to support the project needs as well as how process safety requirements will be communicated to the design team in a timely fashion all throughout the design phase is imperative to a successful project delivery. In a fast-track project, the power of effective and redundant communication with the design team is of utmost importance.

Specific Considerations for Modular Design

Modular design for offshore facilities is typical. More recently, modular design has also been adopted in onshore setting for the benefits it brings during construction. Due to the generally higher equipment densities in the module, some process safety features for onshore modular designs would be similar to offshore designs.

The risk management framework as depicted in **Figure 2** still applies. However, there are some risk reduction measures which are not typically applied in onshore facilities but may be considered in a modular design. Examples of measures that should be considered are as follows:

- Automation of firefighting and response considering less dependence on the fire brigade due to difficulty of accessing equipment within the module;
- More extensive passive fireproofing;
- Robust isolation, depressuring / de-inventorying and sectioning will be required;
- Vertical segregation may need to be considered;
- Protection measures for electrical and instrumentation rooms located in classified areas;
- Increased protection due to placement of pipe racks and equipment over other fire hazard equipment;
- Provision of sufficient means for personnel to escape during emergency event;
- Distance / time of travel to point of safe refuge during emergency egress from a large module assemblies will be a critical parameter, especially if there is a potential toxic gas exposure;
- Increased asset protection to meet client or insurance requirements.

These points are considered during the development of the base design utilizing a modular mid-scale LNG solution. It should also be noted that execution sequence of process safety activities are adjusted to support the module installation schedule. Additional studies which are not typically performed onshore but are performed offshore may need to be performed (i.e. dropped object study, escape/muster analysis, etc.).

2.3 Considerations During the Replication Process

As discussed in Section 1.1, replication of the modular mid-scale LNG is typically done either to increase capacity through the addition of trains at one site or to utilize the base design and install it in a different location. As it is with most copy jobs, there is a misconception that all process safety features of the base design can be carried over in the replicated design which is arranged differently or will be installed in another location or in a different type of facility. In contrary, the fact of the matter is that although the inherent process safety hazards associated with the base design may not change in most of the part of the process units, not all process safety features can be copied as other considerations change along with the change in location, environmental conditions, regulatory requirements, and plot arrangement. On the other hand, one can chose to take the less complicated step which is to start from scratch. However, this will not support the main objective of why a design is being replicated. The actual solution will be somewhere between these extremes.

Risk management during the replication process employs the same framework as presented in **Figure 2**. The main difference is that there will be an existing baseline information which can be used to efficiently complete associated design activities. To facilitate the process of selecting which process safety features can be retained from the base design, efforts should start in recognizing what will be the pertinent differences. This can be achieved by performing a gap analysis of the main drivers to the risk management process. To provide a structured approach in the gap analysis, review can start with the top three (3) drivers which can be categorized into – 1) the minimum mandatory process safety requirements, 2) the applicable risk tolerability criteria and 3) the facility risk profile. **Table 3** provides a summary of where the usual sources of changes are and the impacted activities and deliverables during design phase.

2.3.1 Minimum Process Safety Requirements

Process safety requirements based on local regulations and the list of mandatory industry codes and standards to be applied could differ depending on which part of the world the facility will be built. Though in most cases, the requirements are similar, there could be some differences based on location or client-specific prescriptive requirements. Moreover, for some locations or clients, performance-based approach in risk management or demonstration of risk reduced to as low as reasonably practicable (ALARP) together with the submission of design safety case are required.

2.3.2 Risk Tolerability Criteria

Client risk tolerability criteria typically covers impact to personnel, environment and the asset, is usually presented in form of a risk matrix which is utilized for the analyses such as HAZID, HAZOPs, as well as SIL assessment. In most cases, risk matrices are qualitative, hence, it is inevitable that the team's judgement will be exercised. Decision if additional measures are required is greatly influenced by the company's risk tolerance and culture towards process safety.

Table 3. Main Source of Changes in Replication Process and Their Impacts

Process Safety Drivers	Source of Changes during Replication	Impacted Process Safety – Related Activities / Deliverables
Minimum process safety requirements	Local regulations Applicable codes and standards Client-specific requirements Insurer's requirements, if applicable Certifying body requirements, if applicable	HSE philosophies Engineered risk reduction measures design and their performance requirements
Risk tolerability criteria	Local regulations Client tolerability criteria	HSE philosophies HAZID/ENVID results HSE studies results Engineered risk reduction measures design and their performance requirements
Risk profile	Facility type Site environmental and metreological conditions Process technologies & conditions Plot layout/arrangement	HSE philosophies HAZID/ENVID results HSE studies results Engineered risk reduction measures design and their performance requirements

Semi-quantitative assessments such as layers of protection analysis (LOPA), for example, requires a target mitigated event likelihood (TMEL) which can also differ from one client to another. Risk tolerability criteria being utilized for the quantitative analyses such as fire and explosion risk analysis (FERA) or cryogenic risk analysis (CRA) in most cases, will align with the industry practice, but there are instances wherein client adjusts the criteria to be applied.

Risk tolerability criteria enforced by local regulators is normally in terms of risk to personnel onsite and offsite and will vary in different parts of the world. This criterion is utilized in preparing the quantitative risk analysis (QRA) for LNG facilities which in most cases are required to be submitted to regulators for permitting purposes.

2.3.3 Risk Profile

If the replicated design will be installed on the same site, for example, as a second train for capacity expansion, changes in the plot layout and arrangement may require update of HSE studies for the facility and may impact the engineered risk reduction measures design. To manage this, some clients already include considerations of the future trains in the base design so that minimal work will be required during detailed design of the future trains.

However, if the base design is replicated and will be installed in a different location (i.e. from Asia to North America), or different facility type (i.e. from onshore to offshore), in addition to minimum requirements being different, the risk profile of the facility could significantly change. Process design will need to be revisited to account for site-specific conditions. Environmental and meteorological conditions will require new considerations and can greatly influence the results of the HSE studies.

The plot plan will also most likely be altered which could change how the critical receptors are placed in relation to the sources of hazards therefore, changing the risk reduction measures required. Implementing an execution philosophy that aims to eliminate or minimize modifications within the modules and allowing only changes that are considered mandatory to safety and operability can reduce the overall plot plan changes. Offsite receptors can differ from one site to another. This is not an issue in offshore setting. However, in this facility type, escalation risk and its impact to personnel and asset significantly increases.

Once differences in these main drivers are understood, impact to the process safety features in terms of risk reduction measures provided will need to be confirmed. **Table 4**Error! Not a valid bookmark self-reference. presents a matrix depicting which mitigative measures are potentially impacted if the above discussed factors changed. It should be noted that the table is not all inclusive, but instead, aims to provide a starting point for assessing what could potentially change due to new site location, client, or different facility type.

3.0 Case Studies

The case studies discussed in subsequent subsections offer examples how the recommended practices as discussed in previous sections can be applied. These cases are based on real projects, with some tweaks to provide better demonstration of how process safety features differ as driving factors changes.

3.1 Case Study 1 (Base Design): Modular Mid-Scale LNG on Jack-Up Rigs

The mid-scale LNG reference design chosen utilizes the Chart Integrated Pre-cooled Single Mixed Refrigerant (IPSMR®) liquefaction technology and UOP pre-treatment. The base design is a 1.4 MTPA LNG train which considers US Gulf Coast parameters (i.e. gas composition, inlet gas pressure, etc.) as the process design basis. An onshore reference design was modified for installation on three (3) offshore jack-up rigs (**Figure 3**), each separated 10 meters apart, to be installed in Gulf of Mexico. Two are the process rigs: one rig housed gas treating unit which pre-treat incoming gas from the pipelines; the second rig housed the liquefaction unit and the LNG hose reels used to transfer LNG product to the floating storage units. Both were provided with workshops, local electrical room and emergency equipment, all installed in spaces within their hulls. The third rig carried the utilities, gas turbine generators, helicopter pad, control building and the living quarters (LQ). It has its own local electrical room which is on its main deck while the emergency equipment and other manned spaces are within its hull.

Table 4. Mitigative Systems Potentially Impacted during Replication Process

	Minimum Process Safety Requirements (1)	Client-specific Requirements (2)	Environmental/ Meteorological Conditions (3)	Process Technology/ Conditions	Plot layout/ arrangement	Facility Type (4)
F&G detection provisions	✓	✓	✓	✓	✓	✓
Ignition control	✓	✓		✓	✓	✓
Passive fire and explosion protection	✓	✓	✓	✓	✓	✓
Cryogenic protection		✓	✓	✓	✓	✓
Active fire protection system	✓	✓	✓	✓	✓	✓
Emergency Shutdown system		✓		✓	✓	✓
Relief Depressurization, and Flare system		✓	✓	✓	✓	✓
Drainage arrangement	✓	✓		✓	✓	✓
EER arrangement	✓	✓	✓	✓	✓	✓
Lifesaving/ Safety equipment	✓	✓	✓	✓	✓	✓
Environmental protection	✓		✓	✓		✓

Notes:
(1) Local regulations or codes and standards prescriptive requirements can vary. Some locations or standards apply performance-based approach.
(2) Client-specific requirements vary depending on their company's risk tolerance and established standards.
(3) Environmental conditions include meteorological data e.g. wind rose data, climate, ambient conditions, etc. applicable for the location of the project.
(4) Land based or marine/offshore.

Minimum safety requirements stipulated by local regulations and classification society requirements were reviewed and incorporated in the design, where applicable. Approved list of codes and standards for the facility were also considered in design finalization.

Safety critical equipment was protected from major accident events either by location (most safety critical elements are located within the hull separated from the process area by a steel deck), by design, or by provision of active or passive protection. Plot arrangement gave priority to means for personnel escape, evacuate and muster to safe location. A rigorous review of the 3D model was performed during design phase to ensure egress is provided as appropriate.

Figure 3. Photo of an Offshore Jack-Up Rig

One thing to note for this case study is that the design stages presented in Section 2.2 were not employed. All activities as listed in **Table 5** were completed during detailed design stage.

Table 5. Base Design Risk Management Activities

RMP STEP		DETAILED DESIGN ACTIVITIES
Communic ation and	Establish the Context	Review of local regulations and classification requirements
		Establish project requirements
	Risk Assessment	Identify hazards associated with the process technology
		Conduct HAZID workshop

		Perform full suite of offshore HSE studies, including cryogenic risk assessment Perform Detailed HAZOP and SIL assignment including for vendor packages
	Risk Treatment	Develop and finalize HSE philosophies Identify engineered risk reduction measures Prepare and finalize specification and detailed design of engineered risk reduction measures Develop plot layout for construction Finalize detailed customized design
	Monitoring and Review	Perform inter-disciplinary checks Participate in design reviews/workshops Implement management of change system Undertake HSE verification activities

3.2 Case Study 2: Modular Mid-Scale LNG on Fixed Jacket Platforms

The base design was then replicated for installation on three (3) fixed jacket platforms (**Figure 4**) separated 80 m apart to be situated in US Gulf Coast. There were no changes in the capacity of the LNG trains from the Case Study 1. However, incoming gas composition changed, which initiated the need to review the upstream process design prior to gas treating. Facilities to be installed on each platform are the same for that in Case Study 1 (i.e. two process platforms, with one platform for utilities, LQ, etc). Meteorological conditions in the new location are more severe than that utilized for the Case Study 1 which called for the review of the structural design.

Plot layout changes within the gas treating and liquefaction process modules are minimal. Therefore, the process safety features within the modules were essentially carried over from Case Study 1. The off-module arrangement for the topsides process equipment also had minimal changes but interfaces between on module to off-module elements were still reviewed.

More significant changes occurred for the location of flares, associated knock-out drums, workshops, utilities, LERs, and critical safety equipment due to the use of jacket platforms in Case Study 2. These are located in the lower cellar deck which is open versus in Case Study 1 which had enclosed hulls. Moreover, the upper deck is smaller than the lower cellar deck (e.g., similar to tiered-cake arrangement), leaving lower cellar deck equipment and buildings susceptible to effects of accidental scenarios from the process equipment on the upper deck. Similarly, escape and mustering arrangements were also impacted by the change in deck arrangements and had to be fully reviewed.

Availability of the Case Study 1 design enabled purchasing of long lead equipment at the onset of work. To support procurement efforts earlier in the project, design accidental loads for critical receptors, such as firewater pumps and their diesel driven generators, enclosed secondary muster areas, and emergency diesel generators, were specified conservatively accounting for worse case scenarios while HSE studies are still being performed in parallel. Since Case Study 1 risk profile is comparable, results from Case Study 1 HSE studies were used as a reference for Case Study 2 design accident loads.

Figure 4. Photo of an Offshore Fixed Platforms

The risk management process was initiated by performing a gap analysis to understand the differences in drivers and factors to be considered from the Case Study 1, findings are summarized in **Table 6.**

Table 7 presents the risk reduction features which required modification due to gap analysis results. Most were brought about by the differences in plot layout, regulatory requirements, and site and meteorological conditions. Compliance with the local regulatory requirements necessitates some changes to the design. The major design update was associated with the stricter environmental requirements.

Changes in meteorological conditions impacted results of the HSE studies which generated different protection requirements for Case Study 2 such as height and location of flare boom and protection requirements against unignited cold vapor releases. Changes in incoming gas composition parameters were not shown to change any mitigating measures since Case Study 1 features were still considered sufficient. However, process design was modified to address the new conditions (e.g., use of piping with higher pressure rating).

Changes on the plot arrangement extensively impacted several mitigating measures. Therefore, the design of protective systems were revisited to consider the plot plan changes. All HSE studies were re-performed.

Table 6. Summary of Case Study 2 Changes from Case Study 1 – Base Design

Process Safety Drivers	Source of Changes during Replication	Any changes from Case 1?	Impacted Process Safety – Related Activities / Deliverables
Minimum process safety requirements	**Local regulations**	Y(1)	**HSE philosophies** **Engineered risk reduction measures design and their performance requirements**
	Applicable codes and standards	N	
	Client-specific requirements	N(2)	
	Classification/certifying body requirements	Y(3)	**Engineered risk reduction measures design and their performance requirements**
Risk tolerability criteria	Local regulations	N(4)	
	Client tolerability criteria	N	
Risk profile	**Facility type**	Y(5)	**HSE philosophies**
	Site environmental and meteorological conditions	Y(1)	**HAZID/ENVID results** **HSE studies results** **Engineered risk reduction measures design and their performance requirements**
	Process technologies & conditions	Y(6)	**HSE studies results**
	Plot layout/arrangement	Y(7)	**HAZID/ENVID results** **HSE studies results** **Engineered risk reduction measures design and their performance requirements**

Notes:
(1) Due to change in location.
(2) Client remains the same.
(3) No classification required, only certification based on applicable codes and standards.
(4) No prescriptive requirement on risk tolerability criteria. Risk criteria used for both cases are based on what is typically used in the industry
(5) From jack-up rig to fixed jacket platform
(6) Different conditions upstream gas treating
(7) Significant changes off-module. Changes on modules are minimized.

Table 7. Summary of Changes on Mitigating Measures from Case 1 to Case 2

	Minimum Process Safety Requirements	Client-specific Requirements	Environmental/ Meteorological Conditions	Process Technology/ Conditions	Plot layout / arrangement	Facility Type
F&G detection provisions					✓	
Ignition control					✓	
Passive fire and explosion protection					✓	✓
Cryogenic protection			✓		✓	✓
Active fire protection system					✓	✓
Emergency Shutdown system						
Relief Depressurization, and Flare system			✓		✓	
Drainage arrangement					✓	✓
EER arrangement	✓				✓	✓
Lifesaving/ Safety equipment					✓	
Environmental protection requirements	✓					

3.3 Case Study 3: Modular Mid-Scale LNG on a Cylindrical Floater Unit

For Case Study 3, all units from the three jack-up rig base design were to be installed on a single cylindrical floater unit (example is shown in **Figure 5**). The planned location for this was the same location as for the Case Study 1 base design. Therefore, the applicable regulatory requirements, as well as meteorological and environmental conditions remained the same. There were no changes in the LNG train capacity as well as incoming gas composition and operating conditions.

There were additional design considerations for motion during normal operations for a floater. This resulted to some equipment size changes within the modules to account for motion criteria. Therefore, process safety features within the modules along with their interfaces with off-module elements copied from Case Study 1 base design had to undergo an additional gap review.

The most impactful change was the off-module equipment arrangement and location with respect to modules. Separation distances between major process units, gas treating and liquefaction units (including LNG hose reels), utilized for Case Study 1 could not be carried over due to all equipment from three (3) separate triangular rigs being located on one cylindrical floater.

The living quarters, control building, helicopter pad and lifeboats were located on the same level as the process area but were protected with a blast wall. Other units such as riser bay, utilities, other manned spaces, and other safety critical equipment were located in different spaces within the hull separated by fire (and blast, for some) rated walls and steel deck (from process areas) as required by applicable standards and risk assessment.

Long lead items needed to be purchased at the beginning of the project. To be able to support the procurement efforts earlier in the project, preliminary studies need to be undertaken to provide a fit-for-purpose design accidental loads for the critical receptors. Case Study 1 results cannot be used as a reference to select conservative values for specification since the risk profile for Case Study 3 has changed considerably.

Figure 5. Photo of a Cylindrical Floater

The risk management process was initiated by performing a gap analysis from Case Study 1. **Table 8** summarizes the differences in risk reduction drivers for Case Study 3 as compared to Case Study 1.

Table 8. Summary of Case 3 Changes from Case 1 – Base Design

Process Safety Drivers	Source of Changes during Replication	Any changes from Case 1?	Impacted Process Safety – Related Activities / Deliverables
Minimum process safety requirements	Local regulations	N(1)	
	Applicable codes and standards	N	
	Client-specific requirements	N(2)	
	Classification/certifying body requirements	N(3)	
Risk tolerability criteria	Local regulations	N(1)	
	Client tolerability criteria	Y(4)	**HSE studies results** **Engineered risk reduction measures design and their performance requirements**
Risk profile	**Facility type**	Y(5)	**HAZID/ENVID results** **HSE studies results** **Engineered risk reduction measures design and their performance requirements**
	Site environmental and meteorological conditions	N(1)	
	Process technologies & conditions	N(6)	
	Plot layout/arrangement	Y(7)	**HAZID/ENVID results** **HSE studies results** **Engineered risk reduction measures design and their performance requirements**

Notes:
(1) Same location
(2) Client remains the same.
(3) Same classification body and scope.
(4) Same Client but revisited tolerability criteria used for risk studies.
(5) From jack-up rig to cylindrical floater unit
(6) Same conditions upstream gas treating
(7) Significant changes off-module and few changes on module

As depicted on **Table 8**, most changes in process safety features were brought about by differences in plot layout and facility type. Bringing the units from the three rigs together in one facility changed the plot arrangement and as a result, altered the facility risk profile. This consequently changed the facility process safety features. Although the facility owner did not change, the criteria for determining the requirement for additional fire, blast, and cryogenic protection needed to be further defined. This includes re-defining the main safety functions and putting focus on protecting them when determining the protection requirements. Results from the HSE studies required a more stringent protection for some safety critical receptors.

Change in the type of facility from jack-up rigs to Sevan unit introduced additional changes in safety features particularly in drainage arrangement and location of lifeboats and life rafts. Plot plan overhaul required re-performance of all HSE studies and a full review of the safety features to be provided for the facility.

Table 9. Summary of Changes on Mitigating Measures from Case 1 to Case 3

	Minimum Process Safety Requirements	Client-specific Requirements	Environmental/ Meteorological Conditions	Process Technology/ Conditions	Plot layout / arrangement	Facility Type
F&G detection provisions					✓	✓
Ignition control					✓	
Passive fire and explosion protection		✓			✓	
Cryogenic protection		✓			✓	✓
Active fire protection system					✓	
Emergency Shutdown system					✓	
Relief Depressurization, and Flare system					✓	
Drainage arrangement					✓	✓
EER arrangement					✓	✓
Lifesaving/ Safety equipment					✓	✓
Environmental protection requirements						

3.4 Case Study 4: Modular Mid-Scale LNG Onshore

On to Case Study 4 where the base design was to be installed onshore in British Columbia, Canada for a different Client. Since the desired capacity and liquefaction technology to be used are similar to Case Study 1, it was utilized as base design instead of the original onshore reference design. This was to leverage the advancement of detailed design for the modules. Two 1.4 MTPA trains were to be installed in a phased manner (i.e. phase 1 and phase 2). The incoming gas composition and conditions changed. Full containment LNG storage tanks and marine facilities, such as jetty loading facilities, were also installed as part of the project.

Figure 6. Photo of an Onshore LNG Terminal

Due to change in location, regulatory requirements applicable, as well as meteorological and environmental conditions together with the applicable codes and standards required detailed review. The Client provided a full suite of company-specific design specifications which were required to be implemented for the facility.

The plot arrangement changed extensively. Having the facility built on land meant that sufficient separation distance between modules to reduce the risk of event escalation could be achieved. Occupied buildings were located remote from the process area. Opportunity to locate safety critical equipment and buildings away from hazardous area were maximized to eliminate the need to design them against accidental events. These layout inherently safer design considerations changed the protection requirements for this case study.

Similar to above case studies, the risk management process was initiated by performing a gap analysis against Case Study 1. **Table 10** summarizes the changes in drivers for Case Study 4 as compared to Case Study 1.

Table 10. Summary of Case 4 Changes from Case 1 – Base Design

Process Safety Drivers	Source of Changes during Replication	Any changes from Case 1?	Impacted Process Safety – Related Activities / Deliverables
Minimum process safety requirements	**Local regulations**	Y(1)	**HSE philosophies** **HAZID/ENVID results** **HSE studies results** **Engineered risk reduction measures design and their performance requirements**
	Applicable codes and standards	Y	
	Client-specific requirements	Y(2)	
	Classification/certifying body requirements	Y(3)	
Risk tolerability criteria	**Local regulations**	Y(1)	
	Client tolerability criteria	Y(2)	
Risk profile	**Facility type**	Y(4)	
	Site environmental and meteorological conditions	Y(1)	
	Process technologies & conditions	Y(5)	
	Plot layout/arrangement	Y(6)	

Notes:
(1) Change in location.
(2) Different Client.
(3) Verification body as required by local authorities and Client.
(4) From jack-up rig to onshore facility
(5) Different conditions upstream gas treating
(6) Significant changes in plot arrangement

As depicted in **Table 11**, most changes in process safety features were brought about by differences in client-specific requirements, plot layout, and facility type (onshore versus offshore). Operating companies can have different requirements and risk tolerability criteria. Some clients require company-specific design specifications to be incorporated in the design in addition to regulatory requirements. Review of Case Study 1 design basis against client specifications had to be undertaken to determine which features can remain and which need to be modified.

Full suite of HSE studies as required by local regulations and the client were performed with results being vastly different than for Case Study 1. This necessitates the need to modify a number of risk reduction features, including that within the module. Offsite risk also becomes a concern that had to be address in Case Study 4, as well as nearby environmental receptors which can be impacted by the facility operation. Additional measures to meet the offsite and environmental tolerability criteria must be incorporated in the design.

Table 11. Summary of Changes on Mitigating Measures from Case 1 to Case 4

	Minimum Process Safety Requirements	Client-specific Requirements	Environmental/ Meteorological Conditions	Process Technology/ Conditions	Plot layout / arrangement	Facility Type
F&G detection provisions		✓	✓		✓	✓
Ignition control		✓			✓	
Passive fire and explosion protection		✓	✓		✓	✓
Cryogenic protection		✓	✓		✓	✓
Active fire protection system	✓	✓	✓		✓	✓
Emergency Shutdown system		✓			✓	✓
Relief Depressurization, and Flare system		✓	✓		✓	✓
Drainage arrangement		✓			✓	✓
EER arrangement		✓			✓	✓
Lifesaving/ Safety equipment		✓			✓	✓
Environmental protection requirements	✓					✓

4.0 Conclusions and Recommendations

As noted by Patel et al. (2021), one particular benefit of modular mid-scale LNG solution and the reference design work is the potential reduction in the time needed to perform conceptual studies, front-end engineering design, and time needed to reach approval for the execution of a project. Reference designs are available as a starting point which can then be modified to meet specific requirements for a certain client. Once reference design is customized and made client-specific, it can then be replicated with the purpose of meeting the total plant capacity which in conventional practice would be achieved by a single large-scale plant. The customized base design can also be replicated with the intention to install the modules in other locations or another type of facility (e.g. land or marine based).

During the development of the reference design, hazard identification is initiated by the review of HMB, PFDs as well as safety datasheets on involved substances. Safety features integral to process design such as implementation of ISD concepts, basic process control systems and typical safety instrumented functions are incorporated during P&ID development and equipment specification. Due to several factors influencing the risk management process, there is a limited process safety work that can be done during reference design development.

Replication of the modular mid-scale LNG solution on the same site can be straightforward and installation of the future trains can be considered in the base design to minimize detailed design work in the future. The challenge comes when the base design is to be replicated and installed in a different location or facility type or for a different client as illustrated by the case studies. To facilitate the process of selecting which process safety features can be carried over from the base design, efforts should start in understanding what the pertinent differences are. This can be achieved by performing a gap analysis of the main drivers of the risk management process. To provide a structured approach in the gap analysis, review can start with the top three (3) drivers which can be categorized into – 1) the minimum mandatory process safety requirements, 2) the applicable risk tolerability criteria, and 3) the facility risk profile. Once changes to these main drivers are understood, the impact on the process safety features in terms of risk reduction measures provided will need to be confirmed.

Utilizing the gap analysis results as the basis in determining which process safety features can be copied during the replication process provides a solid basis for justifying why some features can remain as is, and more importantly, why some need to be modified. This allows the project team to streamline the work process and assign resources where they are truly needed to support the aggressive schedule without compromising quality of the outcome of process safety activities.

Finally, application of the risk management framework together with the gap analysis, assists the project team in maintaining the focus in working towards the goal of delivering compliant and safe design to our clients who will ultimately operate the facility.

5.0 References

1 BARKER, S., and LUNA, V. (2023) ***Process Safety Considerations in Designing Modular Mid-scale LNG Facility***. In Fluor Global HSE Engineering and Design Virtual Conference 2023. 16th and 17th of May 2023. Virtual.

2 INTERNATIONAL ASSOCIATION OF OIL AND GAS PRODUCERS (2018) ***Asset integrity – the key to managing major incident risks.*** [Online] England: IOGP. Available from: Asset integrity – the key to managing major incident risks | IOGP Publications library. [Accessed: 05/16/2023]

3 PATEL, D., MARAMBA T., and BORSOS, S. (2021) 'Green' Modular Mid-Scale LNG Liquefaction. In ***Gastech Technical Conference 2021***. 21st to 23rd September 2021. Dubai, UAE.

SPRING24
+20TH GCPS
A Joint AIChE and CCPS Meeting

CONSIDER CLIMATE CHANGE IMPACTS IN YOUR NEXT HAZARD ANALYSIS: THE NEED OF A NEW PERSPECTIVE TO REVIEW NATURAL HAZARDS IN CHEMICAL PROCESS FACILITIES.

Marco-Antonio Medrano-González
RSC. Operation Manager
Zacatecas 70, 309. Col. Roma Norte, Mexico City
marco.medrano@risksc.com

Prepared for Presentation at
American Institute of Chemical Engineers
2024 Spring Meeting and 20th Global Congress on Process Safety
New Orleans, LA
March 24-28, 2024

CONSIDER CLIMATE CHANGE IMPACTS IN YOUR NEXT HAZARD ANALYSIS: THE NEED OF A NEW PERSPECTIVE TO REVIEW NATURAL HAZARDS IN CHEMICAL PROCESS FACILITIES.

Marco-Antonio Medrano-González
RSC. Operation Manager
Zacatecas 70, 309. Col. Roma Norte, Mexico City
marco.medrano@risksc.com

Keywords: Climate change, Climatic Risk, Climatic Vulnerability, Facility Siting Studies, Global Warming, HAZID, HAZOP, Mind Map, Natural hazards, What-if? Analysis, Weather.

Abstract

There is no definitive evidence of how climate change impacts the safety of chemical process facilities. Particularly the oil & gas sector, recognizes that climate change is a relevant factor affecting its onshore and offshore operations. In our experience leading and revalidating hazard analysis, there is a generalized perception that natural hazards have increased their intensities and frequency of occurrence. In other cases, relatively new natural hazards with the potential to damage asset and personnel, interrupt operations, or delay logistics movements have begun to appear in geographic areas where these hazards were not present before.

It is reviewed what climate change is and how it can be a causal factor of some natural hazards. An initial Mind Maps is developed to summarize how those factors related to climate change can be correlated as causes of natural hazards.

An initial Mind Map on Climate Change has been developed to complement and guide brainstorming sessions of HAZID, What if? and HAZOP Analysis. Some cases of impacts of natural hazards along the oil and gas supply chain are briefly described.

1 Introduction.

In our experience leading and revalidating hazard analysis, a generalized perception has emerged in recent years on the behavior patterns modification of some natural events related mainly to weather in a geographic region. Four generic perceptions have been found where variations in the behavior patterns of specific weather events have appeared. These differences have been recognized with the ability to affect the process safety of a chemical process facility:

a) Increasing the severity or intensity level.
b) Increasing frequency of occurrence.
c) Variations in the seasonality of weather events.
d) New natural hazards are beginning to be presented in geographic regions where they did not present before.

There is no definitive evidence about how climate change impacts the safety of chemical process facilities. Some figures and statistics suggest a continuous occurrence of abnormal natural hazards triggered by climate change in different regions of the globe [4].

Climate change is a more complex topic than what is described in this paper and the development of many natural events has multifactorial causes, so the use of Mind Maps proposed in this work should be applied as a basic and initial guide to prepare more organized hazard analysis

2 Basic Definitions of Climate, Weather, Climate Change, Global Warming and Climatic Vulnerability.

Five concepts related to climate change are used interchangeably, but they refer to different situations.

Climate corresponds to atmospheric patterns over long periods of time, usually 30 years or more of registered data. Different regions of the world have different climate, for example hot and humid, rainy, cold and windy, what is called region´s climate systems.

Weather refers to the actual condition of the atmosphere in a definite moment and site. Weather can change hour-to-hour, day-to-day, or month-to-month. The region's weather patterns tracked for at least 30 years are considered climate [4].

Climate does not change as often and suddenly as weather does.

Climate change has changed during the earth´s history, but this change has occurred gradually and slowly over hundreds or even thousands of years by natural factors [5]. Main natural factors that modify climate are the following:

- Movement of tectonic plates.
- Volcanic activity.
- Tilt of Earth´s axis.
- Periodic variations in sunlight intensity.

At least since the 19th century, human activities have emerged as a new factor contributing to the modification of climate in an accelerated way. Human activities have provoked significant variations in the weather patterns of several regions of the world. It has been documented how human activities have contributed to the modification of climate for over a century in different ways, such as rising temperatures, melting glaciers, and other weird and slight perceptible modifications in the normal behavior patterns of the weather. The planet´s temperature has risen about 1.1 °C (2 °F) since the end of 19th century considering the Industrial Revolution as the start point of the climate change from human activities until now [5]. The atmosphere is the most variable part of a region´s climate systems, so we can conclude that human activity has started to alter the atmosphere.

A more formal definition of climate change used by the Intergovernmental Panel on Climate Change (IPCC) refers to a change in the state of the climate that can be identified by changes in the mean and/or the variability of its properties, and that persists for an extended period, typically decades or longer. This definition refers to any change in climate over time, whether due to natural variability or as a result of human activity [17].

Global Warming is the accumulation into the atmosphere of greenhouse effect gases such as carbon dioxide, methane, nitrous oxide and halogens gases and others gases, coming from different human activities in an unprecedent rate emission [5]. The emission of these gases in the last 150 years has gradually altered the movement patterns and chemical composition of the global atmosphere.

Climatic vulnerability is the propensity or the susceptibility of a geographic region where is located a chemical process facility that can be highly, moderately or slightly impacted by Climate Change. Climate Change can increase or decrease the vulnerability of a chemical process facility.

3 A look to the Oil & Gas Industry.

Any part of the oil and gas supply chain can be vulnerable to potential impacts from natural hazards triggered by Climate Change on its assets, operations and logistic movements at different regions of the globe [18]. The supply chain of the oil and gas industry is divided into three general sectors:

- Upstream involves gas and oil exploration drilling and oil and gas production activities and their transportation until processing centers such as refineries.
- Middlestream involves processing of oil and gas in refineries and the storage, transportation and distribution systems of the different refined products.
- Downstream involves storage and distribution of refined products to final consumers.

Any facility in any of these sectors involves potential high-risk profile operations and activities handling and processing chemical hazardous substances that are especially vulnerable to any variation in intensity or frequency of occurrence of any natural hazards such as flooding, landslide, lighting, flash floods, ice storms, extremely high winds, or extreme environmental temperatures, among others.

Facilities in regions frequently subjected to harsh weather or exposed to excessive temperatures are more vulnerable. A new risk term related to climate change (climate or climatic risk) impacting the overall safety of a chemical process facility should be considered to address hazard analysis to understand how to reduce conditions of potential future accidents.

4 Chemical Process Facilities also contribute to Climate Change.

Onshore and offshore operations of the oil and gas industry around all the world generate and emit high amounts of greenhouse gases into the atmosphere. Paradoxically, its production operations and logistic activities are also vulnerable to the effects of sudden, unpredictable and drastic changes in weather conditions. It is estimated that two-thirds of global greenhouse emissions are attributable to the Oil & Gas Industry [5].

The oil and gas industry contributes to global warming with the emission into the atmosphere of two main types of greenhouse gases: 1) combustion gases such as carbon dioxide and nitrous oxides produced from the combustion of natural gas or refined oil products such as gasoline, diesel and jet fuel and 2) unburned gases, mainly methane, from extraction, production or processing operations that are flaring or venting directly to the atmosphere.

There are many studies on the contribution to global warming from oil and gas industry activities, but little assessment of the impacts of climate change along the oil and gas supply chain. In our own experience leading and revalidating hazard analysis studies in several types of facilities of this industry, climate change begins playing an important role, affecting the efficiency of onshore and offshore operations as well as providing conditions for future accidents. This situation is not exclusive of the oil and gas industry, it is extensive to other chemical process facilities in other industrial sectors.

5 What is a Mind Map and how use it in Hazard Analysis.

A form to take into account contemporary environmental risk in hazard analysis, is the use of mind maps which is a graph model that links and arranges concepts, ideas, data, and other relevant components of a problem under analysis in a form similar to how the human brain arranges concepts into related branches. Three elements make up a mind map:

1) A core or central idea of the concept to be reviewed.
2) Branches that connect the central idea with other topics or basic ideas linked to the core idea.
3) Other secondary ideas can be linked with the basic idea through one or more branches in a third level. Other levels may be joined and so on, yielding a tree.

A mind map represents dynamic interrelationships around a main concept or central idea to be analyzed. A mind map has been developed in this work showing how some factors related to climate change can qualitatively be correlated as a cause of a natural hazard or can also be a consequence. The mind map can be helpful in stimulating or guiding efficiently the brainstorming sessions during hazard analysis. The initial mind map developed to review climate change impacts on the process safety of a chemical facility is shown in the next figure.

This mind map can be used as a complementary or guiding tool in HAZID, What-if? or HAZOP Analyses showing a catalog of several types of natural hazards.

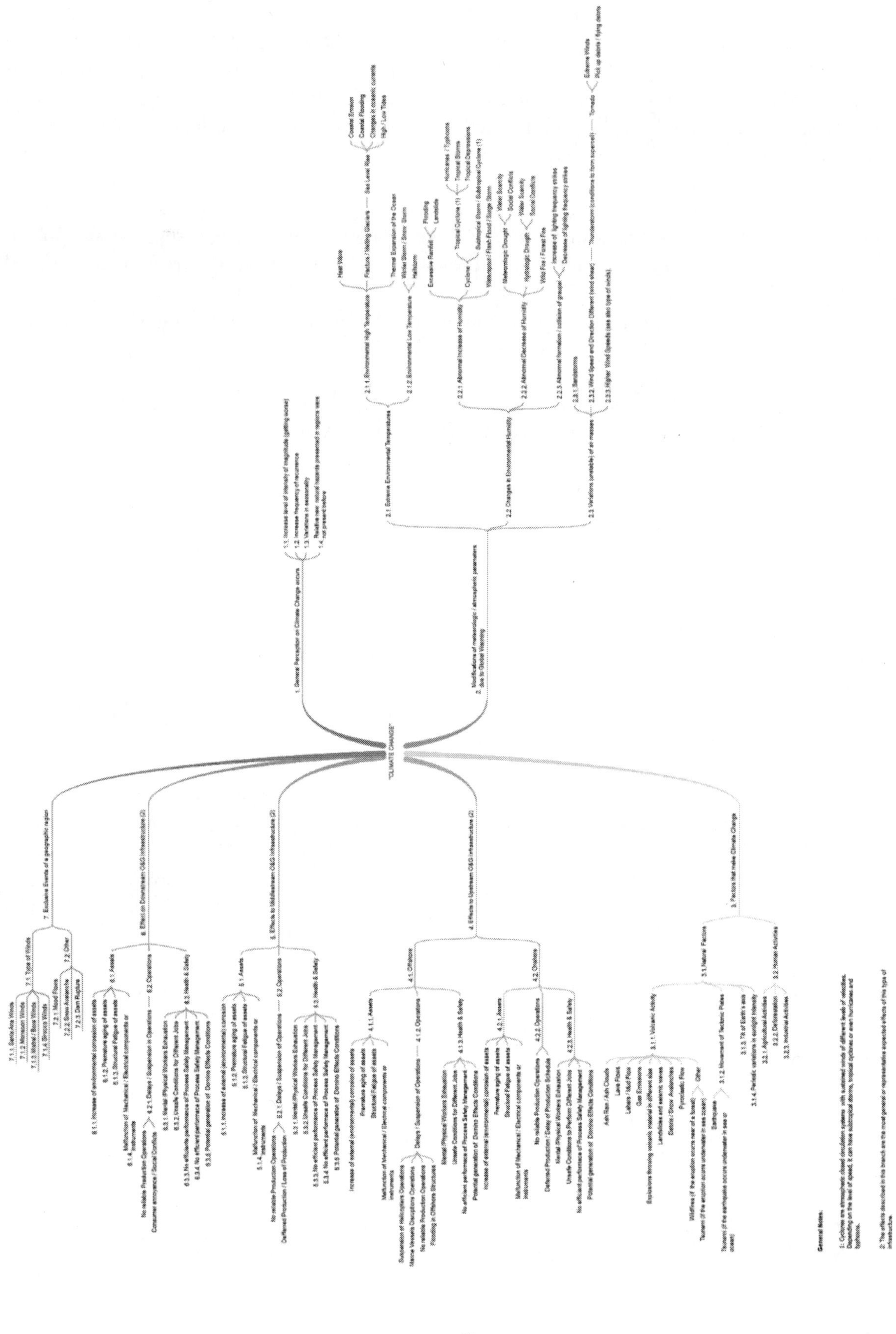

Figure 1. A Mind Map as an initial description of Natural Hazards triggered by Climate Change.

5.1 Reviewing Climate Change with HAZID Analysis.

HAZID reviews general hazards or makes a preliminary hazard analysis of a chemical process facility. It applies an initial checklist of major concepts or Hazard Categories to be reviewed. For each Hazard Category, one or several guidewords are related. A guideword is a particular or representative hazard of a Hazard Category.

A Hazard Category can be a Natural Hazard or apply words such as climate change or climatic risks directly. For this Hazard Category, several natural phenomena can relate guidewords as:

- Earthquake / Tsunami,
- Flash flood,
- Hailing,
- Hurricane / Typhoon,
- Lighting,
- Rainfall,
- Seal Level Rise,
- Volcanic Activity, etc.

Depending on the situation, climate change can be used as a cause of any of these guidewords.

5.2 Reviewing Climate Change with What-if? Analysis.

What-if? Analysis formulates hypothetical questions beginning with the question word "what-if...?". It is easily adaptable to any circumstance to analyze and not exclusively focused on the chemical process safety. Typical What-if? question could be referred directly to a particular natural hazard, to a point on the general perception of climate change, or a variation in any meteorological or atmospheric parameter. Some examples of What if? questions can be:

- "What if" a new natural hazard (tornado, rainfall, etc.) is presented in the region where the facility is located?
- "What if" a natural hazard (tornado, hurricane, flash flood, etc.) increases the level of its intensity?
- What if an abnormal increase of the environmental humidity occurs?

In some cases, the same What if? question can describe the cause being reviewed but also that cause can be a consequence depending on how the what-if? question is formulated, for example:

- What if a flooding event is presented? – flooding is assumed as a cause.
- What if an excessive rainfall event is presented? - in this case, flooding could be a consequence of excessive rainfall.

What-if? Analysis formulates hypothetical questions beginning with the question word "what-if...?". It is easily adaptable to any circumstance to analyze and not exclusively focused on the

5.3 Reviewing Climate Change with HAZOP Analysis.

HAZOP identifies safety and operability problems that may occur in a chemical process during normal operations, as well as no conventional operations such as start-up or shutdown activities. HAZOP induces hypothetical deviations to operation process parameters using guidewords with particular meanings, inducing a deviation in the normal operation parameters and reviewing what could happen.

Climate change seems not to have a direct and immediate influence on the operation of a chemical process, but can impact the overall performance or the safety performance and operability of the entire process. Climate Change can be applied as a general deviation similar to leak or rupture deviations describing the natural hazard to be analyzed (earthquake, volcanism, hurricane, thunderstorm, etc.). Also, a deviation can be applied more specifically with a guideword such as: High / Low Environmental Temperature, or Abnormal Increase / Decrease of environmental humidity, all due to climate change. or abnormal increase / decrease of environmental humidity.

During a revalidation of a HAZOP study of a gas pipeline, a non-expected winter storm affected the operability and availability of the system, several sections were frozen and the natural gas supply was interrupted for several days. A deviation of Low Environmental Temperature was applied to review what happened to this system.

In a new gasoline and diesel storage and distribution terminal project, a potential risk related to volcanic activity was identified using a mind map. Volcanic ash emission could involve potential plugging of vacuum-pressure valves and mechanical damage (collapse) to the roof of the storage tanks. Plugging of the overall drainage system of the facility was also identified as a risk of concern.

5.4 Natural Hazards Analyzed with Mind Map.

A list describing some natural hazards linked to Climate Change has been prepared. These natural hazards are assumed with the potential to impact the process safety of any chemical process facility [1]. These descriptions intend to help in planning hazard analysis or help to formulate in a better way, questions, causes, or consequences related to impacts from climate change to chemical process facilities.

Coastal Erosion: It is the process by which the sea waves gradually erode the land, causing the coast to be spoiled or worn away. Some factors contributing to coastal erosion are:

1) Waves with large fetch or the maximum length of open water over which the wind can blow unobstructed, creating waves with very high energy levels.

2) Strong winds blowing for a long time create destructive waves, and no buffer or beach areas in the coastline are presented.

3) Impact of waves on soft material cliffs, among others.

Drought: It is a period of drier than normal conditions that results in water shortage problem in a particular geographic region. The amount of annual rainfall precipitation varies year to year in a region but in a particular period, this amount of rainfall reduces drastically. Water flows on the ground decline and the water level in lakes or water levels in water wells decreases considerably. The humidity of the environment decreases as the amount of moisture in soils is reduced [6]. Water scarcity and social conflicts can be relevant consequences.

Earthquake / Tsunami: It is produced by the movement of the earth´s inner plates (tectonic layers). When such plates move, collide with each other or they are deformed creating seismic waves that pass through masses of different geologic structures such as rocks or clays releasing the energy stored in those movements in form of suddenly shaking of the ground [9]. Earthquakes occur with spatial but none of their geologic aspects on magnitude or timing can be predicted.

A particular underwater disturbance associated with an earthquake with epicenter in the ocean floor is the Tsunami. A Tsunami is an unusual series of giant waves with large periods that can quickly flood shores and destroy any coastal infrastructure or community. Another causes of Tsunami can be marine landslides and marine volcanic activity [10].

Earthquakes / Tsunamis and Volcanic activity are assumed as natural factors that contribute to natural climate change. They are not affected in their frequencies of occurrence by weather variations.

Flooding: It is a consequence of significant amounts of water spilling onto typically dry ground and can occur from rivers, lakes, streams, the ocean overflowing or too much rain. Floods can have several natural causes related to periods of extremely large storms, hurricanes / typhoons, river overflows or tsunamis [3].

Hailing: it is pieces of tiny, medium, or large size of ice fall to ground from thunderstorm or other types of clouds formed. Hail is formed when water droplets rise into the atmospheric air with temperatures close to freezing point and these droplets freeze. These frozen droplets are lifted by an updraft into a cloud or thunderstorm. The longer these droplets stay in the freezing air, the more layers of ice are added creating hailstones. When the updraft cannot support the weight of the hailstones formed, they fall to the ground.

Heatwave: It is an unusual period of hot weather with high extreme outside environmental temperatures lasting several days. It results from a mass of trapped air without any movement or flow in a particular geographic zone for a long period of time. Atmospheric high-pressure systems push air downward keeping the warm air close to the ground trapped, avoiding this warm air can rise [7].

Hurricane / Typhoons: Both are tropical cyclones with maximum sustained surface wind around 79 mph or 119 km/hr or more and they occur in different geographic regions with slight differences in intensity levels. A mass of air spiraling around a point of low pressure is called a cyclone. Hurricane term is used for the tropical cyclones formed at the northern Hemisphere and the east of the Greenwich Meridian. Typhoon term is used for the tropical cyclones formed at the Pacific Ocean, at the west of the Greenwich Meridian (between longitude or meridian lines 100E and 180E) and in the northern hemisphere. Typhoons are generally stronger in intensity than hurricanes because western Pacific waters are warmer than the eastern Pacific and Atlantic Oceans [8].

Hurricane intensity is classified into five categories according to the Saffir-Simpson Wind Scale. Typhoons can also be classified on the same Saffir-Simpson Scale but can also be classified according to the Japan Meteorological Agency Scale. Frequency can be different; hurricanes occur10 to 15 per year while typhoons occur 25 to 30 per year [7].

Landslide: It corresponds to the movement of rocks, masses of soil and debris down in a sloped section [3]. Landslides have different causes such as extreme rainfalls, earthquakes and volcanic activity.

Lightning: It is produced by the flow of electric current from positively and negatively charged particles. Approximately at 5000 meters above sea level, small clumps of ice crystals are formed in the clouds and gradually become hail. When hail is formed, it constantly collides and rubs with the ice crystals still forming in the clouds or other hail particles, inducing positive and negative electrical charges. Ice, hail and semi-frozen water drops are important to lighting formation [11] The difference in electrical potential between the charged cloud and the ground produces an electrostatic channel resulting in a sudden electrical energy transfer from the bottom of the cloud toward the ground that it is seen as lighting [11].

Rainfall: It is the total amount of precipitation that falls to the ground surfaces as rain (water droplets from clouds). The moisture, or water vapor of the atmosphere is carried upward into air masses by wind turbulence and convection, where it condenses as clouds. This water vapor is finally released by the clouds and falls as rainfall.

Thunderstorm (electrical storm, lighting storm): It is a violent and brief weather disturbance usually associated with lighting, dense clouds formation, heavy rain or hail formation and any of these disturbances accompanied by strong winds. When warm, humid air masses quickly ascend to colder areas of the atmosphere, they form thunderstorms [12].

Thunderstorms are also a precursor of other meteorological phenomena such as lighting, hailing, flash floods, or tornadoes (a particular thunderstorm creating a supercell is needed to form a potential tornado).

Tornado: It usually begins with a specific thunderstorm called supercell, a type of storm bringing hail, strong winds, lighting and flash floods. A supercell is formed when air masses become unstable due to wind blowing in different velocities, directions and altitudes (wind shear). When wind at ground level blows in one direction and wind in higher altitudes blows in a different direction, they cause a tube of air and the warm air mass rises up (updraft). An updraft can turn in from a horizontal rotating air tube into a vertical one. When this happens, the whole storm begins rotating making a supercell or funnel cloud that extends to the ground, producing winds up to 300 miles per hour [13].

Sea Level Rise: It is the most evident effect of the Global Warming. The increase of environmental temperature has created melting of glaciers, loss of ice sheet in the polar ice as well as the thermal expansion of oceans. The sea level rises about 2 to 3 millimeters every year contributes to more frequent coastal flooding and more severe coastal erosion [14]. Additionally, the sea level rise can alter the climate of a geographic region and make other related weather phenomena more intense in severity.

Volcanism: It is the superficial discharge of molten rocks, pyroclastic fragments, ashes, hot water, and several type of volcanic gases as steam mainly, carbon dioxide, sulfur dioxide, and other gases with corrosive properties in smaller quantities such as hydrogen sulfide, hydrogen chloride and some halogens [15, 16]. Volcanic gases emitted into the atmosphere can modify the climate positively or negatively. Volcanism is assumed to be a natural factor of climate change and is not related directly to human activity.

Certain meteorological phenomena and natural events are unique to a given geographic area and can be impacted by climate change. Certain wind types, such as the Santa Ana, Monsoon, Mistral, Bora, and Sirocco winds, are specific to certain geographic areas and can vary in severity and seasonality due to climate change.

6 Conclusions.

This work is a first approach to review potential impacts of natural hazards triggered by climate change to the process safety of a chemical process facility. Many natural hazard events have multifactorial causes, so the use of mind maps is a qualitative approach (initial guide) managing climatic risk in a more organized way. Climate Change seems to be a modern environmental risk that can promote conditions of future accidents in any chemical process facility. The concept of climate or climatic risk should be addressed in hazard analysis.

Mind maps and other new pedagogical tools must be developed to complement hazard analysis. These tools should continually update as more evidence on climate change is gathered and more information becomes available.

The recommendations obtained from Climate Change in hazard analysis can be of three types:

- Adaptation to new climatic conditions.

- Mitigation of effects or damages from natural events triggered by Climate Change.

- Preparation-response-recovery.

Adaptative recommendations could involve modifications in design criteria of process equipment or develop new specifications of materials or components to resist new harsher environmental conditions. A review of design codes or standards can also be considered as an adaptative measure in order to modify the body of design criteria normally used.

Emergency Response Procedures (ERP) should be reviewed and updated or adapted considering that a natural hazard could increase its intensity, occur more frequently or new natural hazards could appear in geographic areas where these hazards were not present before.

Mitigation recommendations could be related for assessing how new natural hazards can impact the performance of personnel working in more extreme environmental conditions or during drought, heatwave, snow storms, high humidity conditions and other natural events. Human Reliability Analysis studies should consider these new environmental conditions in order to decrease the vulnerability of personnel exposed to extreme weather conditions.

A subtype of mitigation recommendations corresponds to long-term investments in reducing greenhouse gas emissions or actions that save energy or water consumption. These recommendations require a more strategic approach, not only to analyze whether a chemical facility has a high or low climatic vulnerability. To measure how the facility´s operations contribute to climate change is needed the use of environmental performance indicators such as the Sustainability Indexes. These types of indicators can be relevant for this purpose.

Preparation-response-recovery recommendations involve a certain level of planning and some level of reactive actions to face the effects of an eminent natural event. Evacuation procedures, having a backup supply of drinking water and non-perishable food for several days, rebuild facilities after the impact of natural events are examples of these recommendations.

Facility Siting studies for projects of new chemical process facilities should include the potential impact of natural events or hazards triggered by climate change and assume that new extreme weather conditions can be presented in the future operations of the facility. A relevant action is to incorporate from the beginning of these projects, environmental performance indicators such as the sustainable indexes to reduce greenhouse gases emission or look for saving in energy or water consumption.

It is necessary avoid the trap of analyzing speculative scenarios on climate change or scenarios with no clear or strong evidence or no history that a natural hazard or event has occurred in the geographic region of interest before. The value of reviewing climate or climatic risk lies in the fact that hazard analysis is the starting point to make formal risk managing actions.

7 References.

[1] "CCPS Monograph: Assessment of and planning for natural hazards," Center for Chemical Process Safety, [Online]. Available: https://www.aiche.org/sites/default/files/html/536181/NaturalDisaster-CCPSmonograph.html. [Accessed 24 February 2024]

[2] Mandal S, Ramkrishna M. *Semi-quantitative Approaches for Landslide Assessment and Prediction.* Singapore Springer, 2015.

[3] White G, Haas JE, *Assessment of Research on Natural Hazards*. Cambridge, The MIT Press, 1975.

[4] "All about Climate," National Geographic Education, [Online]. Available: https://education.nationalgeographic.org/resource/all-about-climate/. [Accessed 15 December 2023].

[5] "Weather or Climate, What´s the Difference?," National Geographic Education, [Online]. Available: https://education.nationalgeographic.org/resource/weather-or-climate-whats-difference/. [Accessed 29 December 2023].

[6] "What causes Drought?," USGS, [Online]. Available: https://www.usgs.gov/faqs/what-causes-drought. [Accessed 11 January 2024].

[7] "What is Heat Wave?, " NOAA, [Online]. Available: https://scijinks.gov/heat/#:~: [Accessed 19 February 2024].

[8] "Hurricanes vs Typhoon," Diffen, [Online]. Available: https://www.diffen.com/difference/Hurricane_vs_Typhoon#:~:. [Accessed 19 February 2024]

[9]. "Heartquake," Britannica. [Online]. Available: https://www.britannica.com/search?query=earthquake. [Accessed 26 January 2024].

[10] "Tsunami," National Geographic. [Online]. Available: https;//www.nationalgeographicla.com/search?q=tsunami. [Accessed 20 February 2024].

[11] "How Lightning is created," NOAA. [Online]. Available: https://www.noaa.gov/jetstream/lightning/how-lightning-is-created#:~:. [Accessed 20 February 2024].

[12]. "Thunderstorm,"Britannica. [Online]. Available: https://www.britannica.com/science/thunderstorm. [Accessed 27 January 2024].

[13] "How Does Tornadoes Form?," NOAA. [Online]. Available: https://scijinks.gov/tornado/#:~: [Accessed 16 February 2024].

[14] "¿Qué es el aumento del nivel del mar?, National Geopgrahic. [Online]. Available: https//www.nationalgeographic.es/medio-ambiente/que-es-el-aumento-del-nivel-del-mar. [Accessed 19 February 2024].

[15] "Volcanism,". Britannica. [Online]. Available: https://www.britannica.com/science/volcanism. [Accessed 15 February 2024].

[16] "Erupciones Volcánicas: qué son, qué efectos tienen y porque son tan peligrosas", National Geographic. [Online]. Available: https://www.nationalgeographic.com.es/ciencia/que-son-erupciones-volcanicas-que-efectos-tienen-por-que-son-tan-peligrosas_18153. [Accessed 23 February 2024].

[17] "Fact sheet: Climate change science - the status of climate change science today," United Nations Framework Convention on Climate Change. [Online]. Available: https://unfccc.int/files/press/backgrounders/application/pdf/press_factsh_science.pdf. [Accessed 23 February 2024].

[18] Chizaram-Dagogo N. Development of an Integrated Process Safety Management and Climate Change Model for the Oil and Gas Industry. Ph.D. Conventry University, May 2020. [https://://pureportal.coventry.ac.uk/en/studentTheses/development-of-an-integrated-process-safety-management-and-climat. [Accessed 13 September 2022].

AUTHOR INDEX